PHOTONICS IN SPACE

Advanced Photonic Devices and Systems

PHOTONICS IN SPACE

Advanced Photonic Devices and Systems

Caterina Ciminelli
Francesco Dell'Olio
Mario Nicola Armenise
Politecnico di Bari, Italy

W World Scientific

NEW JERSEY · LONDON · SINGAPORE · BEIJING · SHANGHAI · HONG KONG · TAIPEI · CHENNAI · TOKYO

Published by

World Scientific Publishing Co. Pte. Ltd.

5 Toh Tuck Link, Singapore 596224

USA office: 27 Warren Street, Suite 401-402, Hackensack, NJ 07601

UK office: 57 Shelton Street, Covent Garden, London WC2H 9HE

Library of Congress Cataloging-in-Publication Data

Names: Ciminelli, C. (Caterina), author. | Armenise, Mario N., author. |
    Dell'Olio, F. (Francesco), author.
Title: Photonics in space : advanced photonic devices and systems / Caterina Ciminelli
    (Politecnico di Bari, Italy), Mario Nicola Armenise (Politecnico di Bari, Italy) &
    Francesco Dell'Olio (Politecnico di Bari, Italy).
Description: [Hackensack] New Jersey : World Scientific, 2016. |
    Includes bibliographical references and index.
Identifiers: LCCN 2015049067 | ISBN 9789814725101 (hc : alk. paper)
Subjects: LCSH: Space vehicles--Electronic equipment. | Optoelectronic devices.
Classification: LCC TL3000 .C56 2016 | DDC 629.47/4--dc23
LC record available at http://lccn.loc.gov/2015049067

British Library Cataloguing-in-Publication Data
A catalogue record for this book is available from the British Library.

Desk Editor: Suraj Kumar

Typeset by Stallion Press
Email: enquiries@stallionpress.com

Contents

Preface

Photonics is a well-established key enabling technology in several scientific/technological areas such as telecommunications, life science, health care, lighting and so on.

In the last years, photonics is gaining the same role also in space engineering. Optical, opto-electronic, and photonic payloads are currently widely utilized especially for earth observation and several new payloads based on the same technologies are in the development stage. In the satellites currently in orbit, photonic components are included in some sub-systems, i.e. the attitude control and the power supply sub-systems. We expect that in the next decade, photonics will gain an important role in other sub-systems such as the data processing/handling and the communications ones. Also, the optical fiber technology is demonstrating important advantages with respect to the competing technologies in monitoring most critical mechanical structures included in satellites and launchers.

The purpose of the book is to collect and critically review the main results obtained by the scientific community in the field of photonic components and sub-systems for space applications with special emphasis on the integrated micro-photonic devices and photonic integrated circuits.

The book is intended for researchers and Ph.D. students interested in the field of photonics and space engineering. It may also be a useful educational tool in some university-level courses focused on photonics and avionics.

In recent years, the authors have spent an intense research effort on micro-photonic chip-scale gyroscopes and integrated microwave photonics for space applications, working on projects supported by the European

Space Agency and the Italian Space Agency. They use their deep know-how on the potential of photonics in space engineering.

The book includes eight chapters. The first two chapters introduce the topic and briefly summarize the fundamentals of photonic devices. Chapters 3 and 4 are focused on the applications of photonics in the data processing/handling sub-system and the communications sub-system of the satellite platform. Image sensors operating from the ultraviolet to the infrared is the topic of Chapter 5. Chapter 6 reports on the state-of-the-art of photonic gyros, fiber Bragg grating sensors, and optical instruments for space. Solar cells intended for satellites power supply is the topic of Chapter 7. Finally, Chapter 8 briefly discusses some emerging space applications of photonics.

Caterina Ciminelli
Francesco Dell'Olio
Mario Nicola Armenise
Bari, 15th June, 2016

Chapter 1

Introduction

Photonics deals with the generation, manipulation/control, and detection of light waves and photons. Its key enabling potential has been demonstrated for several years in many application areas, such as telecommunications, energy, lightening, environmental monitoring, robotics, industrial production, biomedicine, medical imaging, displays, homeland security, aerospace, defense, and many more, so that its worldwide market is expected to exceed 600 billion Euro in 2020 [1].

In the last decades of the 20th century, photonic devices and optical fibers dramatically improved the performance of the telecommunications systems, also providing the infrastructure for Internet development and its exponential growth in the last few years [2]. In the same years, laser technology has radically modified some of the key manufacturing processes such as welding, cutting, and drilling, also enabling very precise manufacturing techniques at the micro/nanoscale [3]. Photonic technologies, especially optical imaging for intraoperative surgical guidance and medical lasers, are currently widely used in several fields of surgery [4].

The enabling potential of photonics in space engineering has notably grown since the 1960s, when the only photonic devices on satellites were the solar cells. In recent years, photonic components and sub-systems have been crucial for many functions on board a spacecraft, e.g. data handling, attitude and orbit control, as well as strain/thermal mapping. Currently, many payloads for both Earth observation and scientific missions include a wide variety of optical and opto-electronic components, such as lasers, detectors, modulators, lenses, gratings, mirrors, and so on.

1

In this chapter, the three segments comprising a space system are briefly introduced and the criteria for space missions and orbit classification are introduced. Finally, an overview of the space applications of photonics is provided.

1.1 Space system segments

The three segments typically forming a space system are the space segment, the transportation segment, and the ground segment. The space segment includes the payload and the satellite platform carrying it, while the transportation segment provides the transport into space of both the satellite platform and its payload by a launcher. The ground segment controls and monitors the spacecraft and its payload, and processes the data provided by the payload.

The payload is the hearth of the space system, being conceived and developed according to the purpose of the space mission, and its features strongly influence the design of both the satellite platform and the transportation/ground segment. A wide variety of payloads has been developed since the beginning of the space age. The most common payloads are spectrometers, radiometers, magnetometers, cameras, radar systems, transmitters, and atomic clocks for navigation/positioning, transponders for satellite telecommunications, and rover vehicles. The sub-systems forming the satellite platform are listed in Table 1.1, where the basic functionality of each sub-system is mentioned.

The transportation segment includes the launch vehicle, e.g. a rocket such as Atlas V or Ariane 5, equipped with appropriate propellants, and the launch infrastructure. The main components of the launch vehicle are the mechanical structure, propellant tanks, engines, attitude control systems, and flight monitoring systems. In the launcher, the thrust is generated by igniting two propellants, the fuel and the oxidizer, in the thrust chamber. Chemical reactions between the propellants produce very high-temperature gases that expand through a nozzle.

The ground segment includes the control center, where the spacecraft is monitored and controlled, and the ground stations network receiving and transmitting data to/from the spacecraft.

Table 1.1.   Sub-systems forming the satellite platform.

Sub-system	Basic function
Mechanical structure	Accommodate all other sub-systems
Power supply	Generate and efficiently distribute electrical energy within the spacecraft
Thermal sub-system	Keep the temperature of spacecraft components within appropriate intervals
Attitude control	Real-time monitor and control the attitude of the spacecraft in space
Communications	Transmit/receive data (telemetry data, commands, and payload data) to/from the ground stations
Data processing and handling	Process and handle data on board the spacecraft
Propulsion	Change the orbit of the spacecraft

1.2 Classification of orbits and missions

Space missions can be classified according to the area of application, while the orbits are classified according to their shape and the altitude. There is a strong correlation between the target mission of a spacecraft and its orbit.

The vast majority of all the orbits selected for space missions are around the Earth. The low Earth orbit (LEO) is a near-circular orbit with altitude ranging from 300 km to 1,500 km, while the so-called geostationary orbit (GEO) is a near-circular orbit having an altitude of about 36,000 km. The near-circular orbits at intermediate altitudes, from 1,500 km to 36,000 km, are called medium Earth orbits (MEOs). Highly elliptical orbits (HEOs) are elliptical orbits with satellite distance from the Earth at the perigee < 1,000 km and satellite distance from the Earth at the apogee > 36,000 km. The geostationary transfer orbit (GTO) is a highly eccentric orbit in which a satellite is temporarily placed before reaching the GEO orbit. Spacecraft for planetary exploration are placed in interplanetary orbits and their distance from the Earth is up to several billion kilometers.

The most common purposes of the space missions are listed in Table 1.2, with the identification of the orbits typically used.

Table 1.2. Typical purposes of the space missions.

Purpose	Orbit
Earth observation	LEO
Weather monitoring	LEO
Telecommunications	GEO, MEO, or HEO
Navigation	MEO
Astronomy	LEO, HEO
Planetary exploration	Interplanetary orbits
Technology testing	LEO

1.3 Overview of the space applications of photonics

Although some photonic sub-systems for the transportation segment and the ground one have been developed in the last few years, most of the space applications of photonics are relevant to the space segment, i.e. payload and satellite platform.

In some kinds of payload such as cameras and spectrometers, optical and photonic components have been widely used for several decades, while the use of photonic technologies within radar, telecom, and navigation payloads is certainly less mature, in spite of some technological demonstrators having already been developed.

Photonics can play a key role in nearly all sub-systems forming a satellite platform due to its intrinsic advantages with respect to conventional technologies. Opto-electronic gyroscopes have routinely been included in attitude and orbit control systems for several decades, and some important advantages of optical fibers in the implementation of on-board data buses have been proved in several space missions, starting from the 1990s. Since their first utilization in space, in 1958, the solar cells have been included in the power supply sub-system of all satellites. Mapping of strain and temperature in some critical sections of the spacecraft can benefit from fiber Bragg grating technology, whose space applicability has already been proved. Finally, optical wireless links to transfer data from one satellite to another or from a satellite to a ground station have been successfully demonstrated in some recent space missions.

In this last case, the ground station is equipped with an appropriate optical terminal, including many photonic components. This is one of the emerging applications of photonics in the ground segment.

Some launchers are equipped with opto-electronic gyros for attitude control, while the fiber Bragg grating technology seems to be very advantageous to monitor the tanks in some kinds of launcher.

References

[1] Photonics Industry Report 2013. http://www.photonics21.org/.
[2] C. DeCusatis (Ed.) (2013). *Handbook of Fiber Optic Data Communication.* Academic Press, New York.
[3] J. Paulo Davim (Ed.) (2013). *Laser in Manufacturing.* John Wiley & Sons, Hoboken, NJ.
[4] T. Vo-Dinh (Ed.) (2015). *Biomedical Photonics Handbook.* CRC Press, Florida.

Chapter 2

Fundamentals of Photonic Devices

The content of this chapter deals with photonic devices based on optical waveguides. In the last few decades, photonics has been a success story because the large-scale deployment of many photonic devices can be clearly identified in commercial applications. Attention is mainly paid to photonic devices based on optical waveguides, allowing integration of several components on the same chip. Different materials have been experimented depending on the specific applications. The development of both coherent and incoherent light sources has allowed a wide variety of photonic devices to become key elements in different field of applications, such as telecommunication systems, sensing, and signal processing.

After outlining the basic principles of light propagation in dielectric waveguides, this chapter includes the description of some algorithms to carry out the electromagnetic analysis and the design of any kind of dielectric waveguide, the physical behavior and the key parameters of optical microresonators, the principle of operation of photonic crystals (PhCs) and their major applications, the passive waveguide components to be used in integrated photonic circuits, the electro-optic modulators in lithium niobate now commercially available for optical instrumentations and Dense Wavelength Division Multiplexing (DWDM) telecommunication systems. A deep physical insight into the instabilities in optical devices in lithium niobate is also given, due to the large demand for such devices in the photonics market. The chapter also includes a brief description of acousto-optic devices and the basic operation of semiconductor lasers and detectors.

2.1 Maxwell's equations

Light is an electromagnetic wave complying with the laws of electromagnetic theory. Optical frequencies range from the infrared to the ultraviolet in the electromagnetic spectrum.

An electromagnetic wave is formed by two vectors of position and time, the electric field vector E and the magnetic field vector H, which are mutually coupled. Electric (E) and magnetic (H) fields in free space must satisfy the following Maxwell's equations:

$$\nabla \times H = \varepsilon_0 \frac{\partial E}{\partial t}, \tag{2.1}$$

$$\nabla \times E = -\mu_0 \frac{\partial H}{\partial t}, \tag{2.2}$$

$$\nabla \cdot E = 0, \tag{2.3}$$

$$\nabla \cdot H = 0, \tag{2.4}$$

where ε_0 is the electric permittivity and μ_0 the magnetic permeability of free space.

To satisfy Maxwell's equations, E and H must be the solutions of the wave equation, which can be derived from Maxwell's equation:

$$\nabla^2 (E,H) - \frac{1}{c_0^2} \frac{\partial^2 (E,H)}{\partial t^2} = 0, \tag{2.5}$$

where $c_0 = \dfrac{1}{\sqrt{\varepsilon_0 \mu_0}}$ is the speed of the light in free space.

When light propagates in a medium without electric charges and currents, Maxwell's equations are still valid and assume the following expressions:

$$\nabla \times H = \varepsilon_0 \frac{\partial D}{\partial t}, \tag{2.6}$$

$$\nabla \times E = -\frac{\partial B}{\partial t}, \tag{2.7}$$

$$\nabla \cdot D = 0, \tag{2.8}$$

$$\nabla \cdot B = 0, \tag{2.9}$$

where B and D are the magnetic flux density and the electric flux density vectors, respectively. It is well known that while the relation between B and H depends on the magnetic properties of the medium, the relation between D and E depends on the electric properties of the medium:

$$D = \varepsilon_0 E + P, \tag{2.10}$$

$$B = \mu_0 H + \mathbf{M}, \tag{2.11}$$

where P is the polarization density and $\mathbf{M}$ the magnetization density. Of course, P and $\mathbf{M}$ are also related to the electric E and magnetic H fields through the electric and magnetic properties of the medium. In free space $P = \mathbf{M} = 0$, which leads to $D = \varepsilon_0 E$ and $B = \mu_0 H$. In this book, we consider non-magnetic media ($\mathbf{M} = 0$).

A dielectric medium is linear when the vector P is linearly related to the electric field E; it is dispersive if P, at the instant time t, depends on the value of E at the same t and at prior time instants. The medium is homogeneous if the equation relating P to E does not depend on position; the medium is isotropic if P does not depend on the direction of E.

Assuming a linear, homogeneous, non-dispersive, and isotropic dielectric medium, the relation between P and E is given by

$$P = \chi \varepsilon_0 E, \tag{2.12}$$

where χ is the electrical susceptibility. Using this last equation, it follows:

$$D = \varepsilon E, \tag{2.13}$$

with the electric permittivity of the dielectric medium expressed as $\varepsilon = \varepsilon_0(1+\chi)$.

In this case, Maxwell's equations become:

$$\nabla \times H = \varepsilon \frac{\partial E}{\partial t}, \tag{2.14}$$

$$\nabla \times E = -\mu_0 \frac{\partial H}{\partial t}, \tag{2.15}$$

$$\nabla \cdot E = 0, \tag{2.16}$$

$$\nabla \cdot H = 0. \tag{2.17}$$

E and H also have to satisfy the wave equations:

$$\nabla^2 E - \frac{1}{c^2}\frac{\partial^2 E}{\partial t^2} = 0, \tag{2.18}$$

$$\nabla^2 H - \frac{1}{c^2}\frac{\partial^2 H}{\partial t^2} = 0, \tag{2.19}$$

where the speed of light is now $c = \frac{1}{\sqrt{\varepsilon \mu_0}} = \frac{C_0}{n}$, with $n = \sqrt{\frac{\varepsilon}{\varepsilon_0}} = \sqrt{1 + \chi}$ being the refractive index of the medium. In a non-homogenous medium, Eqs. (2.12) and (2.13) are still valid, but ε and χ are now functions of position.

In anisotropic media, P and E are in general not parallel because both depend on the direction of the electric field vector. In this case, with reference to a coordinate axis system (x, y, z), it results in:

$$P_i = \sum_j \varepsilon_0 \chi_{ij} E_j, \tag{2.20}$$

with $i, j = x, y, z$. Moreover, the dielectric properties of the medium are described by the electrical permittivity tensor whose elements are ε_{ij} ($i, j = x, y, z$) and by the susceptibility tensor χ_{ij}. Therefore, an anisotropic, linear, homogeneous, and non-dispersive medium is characterized by nine constants, i.e. the elements of the susceptibility tensor χ_{ij}.

In a dispersive medium, the relation between P and E is a dynamic relation similar to that between input and output in a linear dynamic system. A time delay is also generated. The medium can be described by the response function to the impulse at the input. P vector at time t results from the superposition (if the medium is linear) of the E contributions at all instants of time $\leq t$. The susceptibility of the medium results in a function of the frequency f. This effect, i.e. the dispersion of the susceptibility of a medium, is generated by the fact that the response of the medium to excitation by an optical field does not decay instantaneously. If the material response depends on the resonance frequency ω_0 closest to the optical frequency ω, with a relaxation constant γ, the susceptibility in the time domain is the impulse response of the medium [1]:

$$\chi(t) = \begin{cases} e^{-\gamma t} \sin\omega_0 t & t > 0 \\ 0 & t < 0. \end{cases} \tag{2.21}$$

The condition $\chi(t) = 0$ for $t < 0$ is known as the causality condition, i.e. a medium can respond only after excitation, which is valid for all physical systems. The Fourier transform of Eq. (2.21) shows that the susceptibility has real and imaginary parts, both depending on the frequency.

The relation between P and E is always nonlinear in a nonlinear medium. Although the wave equations (2.18)–(2.19) are not valid, a wave equation for nonlinear media can be derived from Maxwell's equations [2].

Assuming the light as a monochromatic electromagnetic wave, all components of the electric and magnetic fields are harmonic functions of time. Considering the time operator $\exp(j\omega t)$, one can write:

$$E(x, y, z, t) = \text{Re}[E(x, y, z)\exp(j\omega t)], \tag{2.22}$$

$$H(x, y, z, t) = \text{Re}[H(x, y, z)\exp(j\omega t)], \tag{2.23}$$

where $\omega = 2\pi f$ is the angular frequency, and $E(x,y,z)$ and $H(x,y,z)$ are the complex amplitudes of E and H, respectively. Under these conditions, Maxwell's equations can be rewritten as follows:

$$\nabla \times H = j\omega D, \tag{2.24}$$

$$\nabla \times E = j\omega B, \tag{2.25}$$

$$\nabla \cdot D = 0, \tag{2.26}$$

$$\nabla \cdot B = 0. \tag{2.27}$$

Similarly, Eqs. (2.10) and (2.11) can be modified to:

$$D = \varepsilon_0 E + P, \tag{2.28}$$

$$B = \mu_0 H. \tag{2.29}$$

2.2 Phase and group velocity

For a monochromatic electromagnetic wave propagating in the z-direction, the electric field vector can be written as a sinusoidal wave:

$$E(z, t) = A \exp(j\beta t - j\omega t), \tag{2.30}$$

where A is a vector independent of z and t, and β is the propagation constant. The phase of the wave is

$$\varphi = j\beta t - j\omega t, \tag{2.31}$$

and its phase velocity is defined as

$$v_p = \frac{dz}{dt} = \frac{\omega}{\beta}, \tag{2.32}$$

which is the velocity with a point of constant phase.

The phase velocity is a function of the frequency because it depends on the refractive index n, which depends on the frequency. This means that the phase velocity dispersion is due to the fact that $dn/d\omega \neq 0$. This dispersion is called normal when $dn/d\omega > 0$ and anomalous if $dn/d\omega < 0$. However, it is very difficult to have a monochromatic wave. In fact, a wave is usually formed by several frequency components around a center frequency. If we consider a wave packet formed by a number of plane waves and traveling in the z-direction, the resultant wave shows a carrier traveling at the phase velocity and an envelope traveling at the group velocity given by:

$$v_g = \frac{d\omega}{d\beta}. \tag{2.33}$$

This is, of course, the velocity of the wave packet. Moreover, since the energy carried by the wave packet is confined where the amplitude of the envelope is large (we must remember that the energy of a harmonic wave is proportional to the square of its field amplitude), the energy in a wave packet travels at the group velocity. Therefore, while the wavefront travels at the phase velocity, energy and information travel at the group velocity.

An important parameter related to the group velocity is its dispersion represented by

$$\frac{d^2\beta}{d\omega^2} = \left[\frac{dv_g}{d\omega}\right]^{-1} \neq 0. \tag{2.34}$$

The group velocity dispersion can be responsible for the broadening of a pulse and a time delay between different pulses. The sign of $d^2\beta/d\omega^2$ can be either positive or negative. If positive, a pulse with long wavelength travels faster than a short-wavelength pulse, and vice versa in case of negative group velocity dispersion.

Another useful parameter is the so-called group index N, defined as follows. The propagation constant β can be written as

$$\beta = \omega \frac{n(\omega)}{c}, \tag{2.35}$$

v_p is expressed by Eq. (2.32) and

$$v_g = \frac{c}{N}, \tag{2.36}$$

where

$$N = n + \omega \frac{dn}{d\omega} = n - \lambda \frac{dn}{d\lambda}, \tag{2.37}$$

with λ being the wavelength.

2.3 Optical waveguides

Optical waveguides are key components in most opto-electronic and photonic devices for a wide variety of applications.

An optical waveguide is formed by a region with a high refractive index surrounded by regions with a lower refractive index. With reference to the material system utilized, optical waveguides can be classified as step-index (i.e. homogeneous waveguides) and graded-index (inhomogeneous waveguides), or isotropic (isotropic material substrate) and anisotropic waveguides (anisotropic material substrate).

A geometrical classification, based on the number of significant cross-sectional dimensions, includes planar or slab waveguides, which are waveguides having just one transversal significant dimension, and channel or rectangular waveguides and optical fibers, which have two transversal comparable dimensions [3–5].

2.3.1 *Slab waveguides*

The step-index slab waveguide is the simplest kind of optical waveguide. Wave propagation in this waveguide can be discussed either in terms of rays through the eikonal equation or electromagnetic wave theory through the wave equation and Maxwell equations. We refer to this last technique.

Slab waveguides can support both guided and radiated modes. We analyze here the guided modes because they are carrying the energy. Radiated modes are of interest for those components where the light is coupled out of the waveguide and when the loss has to be evaluated in the waveguide.

We describe the transformation matrix method [6, 7], which is useful to analyze any kind of slab waveguide. To this end, we consider a slab, non-homogeneous, anisotropic, and multilayered waveguide, as in Fig. 2.1. This numerical technique does not make any significant physical approximation and provides complete results in a very short CPU time. Of course, the algorithm also remains valid for less complicated waveguide configurations.

A non-homogeneous and anisotropic waveguide can be manufactured using different materials depending on the required optical characteristics. As for the slab waveguides, the most common material is lithium niobate [8–10], due to its own good optical properties. The fabrication technique which gives a good optical waveguide consists of the thermal diffusion of titanium in $LiNbO_3$.

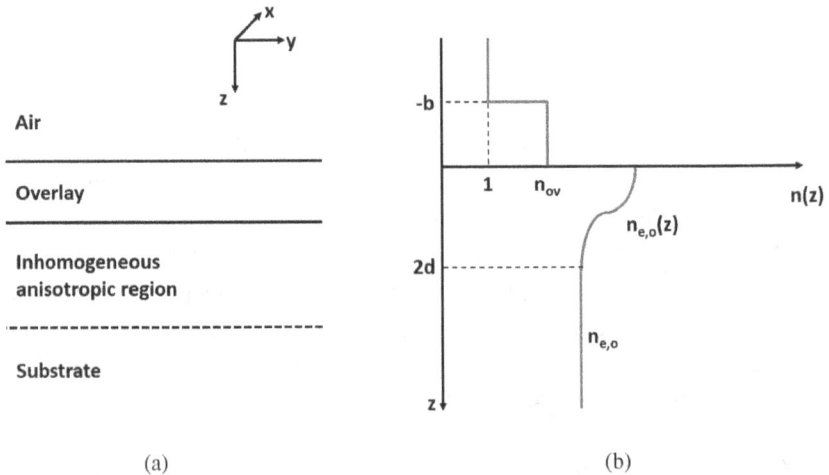

(a)   (b)

Fig. 2.1.   Geometrical (a) and physical (b) configuration of the slab waveguide.

With reference to Fig. 2.1, we assume a Z-cut LiNbO$_3$ crystal, an optic axis c in the z-direction, reference cartesian axis overlapping the principal axes of the crystal, a wave propagation along the x-direction, n_e, n_0 as the extraordinary and ordinary refractive indices, respectively, of the waveguide, n_{es}, n_{os} as the extraordinary and ordinary refractive indices, respectively, of the substrate, n_{ov} as the refractive index of the overlay and d as the diffusion depth. The overlay, also named buffer or loading film, is considered as homogeneous and isotropic, and is responsible for the fine tuning of the propagation constant in the guided layer. We also assume the same profile for the extraordinary and ordinary index in the guide, $n_e(z) = n_0(z)$.

Under these conditions, the permittivity tensor can be expressed by:

$$\tilde{\varepsilon} = \varepsilon_0 \begin{vmatrix} n_0^2 & 0 & 0 \\ 0 & n_0^2 & 0 \\ 0 & 0 & n_e^2 \end{vmatrix}, \tag{2.38}$$

where the equation $\varepsilon_r = n^2$ has been used.

The index profile in the waveguide depends on the fabrication technique. In the case of titanium in-diffusion, when the titanium has completely diffused into the lithium niobate, $n(z)$ can be represented, with good approximation, as a complementary error function, erfc. However, it is important to note that the transformation matrix technique can be applied for any index profile of the waveguide.

We set

$$n_{e,o}(z) = n_{es,os} + \Delta n_{e,o}\,\mathrm{erfc}(z/d), \tag{2.39}$$

with

$$\mathrm{erfc}(z/d) = 1 - \mathrm{erf}(z/d) = 1 - \frac{2}{\sqrt{\pi}} \int_0^{z/d} \exp(-t^2)\,dt, \quad \text{and} \quad \Delta n_e = \Delta n_0 = \Delta n.$$

By considering Maxwell's equations (2.6)–(2.7), if $\mu = \mu_0$ and the time and space operator is $\exp(j\omega t - j\beta x)$, we derive

$$\nabla \times H = j\omega\tilde{\varepsilon}E = j\omega\varepsilon_0\,(n_0^2 E_x i_x + n_0^2 E_y i_y + n_e^2 E_z i_z), \tag{2.40}$$

$$\nabla \times E = j\omega\mu_0 H, \tag{2.41}$$

from which the following two systems, for the transverse electric (TE) and the transverse magnetic (TM) modes, respectively, are obtained:

$$\frac{dH_x}{dz} = j\left(\omega_0 n_0^2 - \beta^2/\omega\mu_0\right)E_y = j\left[\left(k_0^2 n_0^2 - \beta^2\right)/\omega\mu_0\right]E_y$$

$$= j\left(k_0/\eta_0\right)\left(n_o^{\,2} - n_{\text{eff}}\right)^2 E_y \tag{2.42}$$

$$\frac{dE_y}{dz} = -j\omega\mu_0 H_x \tag{2.43}$$

with $k_0^2 = \omega^2\varepsilon_0\mu_0$, $\eta_0 = \sqrt{\mu_0/\varepsilon_0}$, $k_0\eta_0 = \omega\mu_0$, $k_0/\eta_0 = \omega\varepsilon_0$ and

$$\frac{dE_x}{dz} = j\left[\left(k_0\eta_0/n_e^{\,2}\right)\left(n_{\text{eff}}^2 - n_e^{\,2}\right)\right]H_y, \tag{2.44}$$

$$\frac{dH_y}{dz} = -j\omega\varepsilon_0 n_0^2 E_x. \tag{2.45}$$

To solve the TE and TM systems, we introduce the transformation matrix $\Phi(u, 0)$, where $u = k_0 z$

$$F(u) = \Phi(u, 0)F(0). \tag{2.46}$$

This relation means that the electromagnetic field components at $z = 0$ can be related to those at any section of the waveguide through the transformation matrix, whose elements have to be determined, analogously to the matrix representation of electrical quadrupoles. To prove this, we consider the TM modes and set

$$\begin{aligned} F_2 &= E_x, \\ F_3 &= j\eta_0 H_y. \end{aligned} \tag{2.47}$$

So, the TM system becomes

$$\begin{cases} dF_2/du = \left[n_{\text{eff}}^{\,2}/n_e^2 - 1\right]F_3 \\ dF_3/du = n_0^2 F_2. \end{cases} \tag{2.48}$$

At this stage, the matrix elements can be evaluated by solving the TM system formed by two first-order differential equations with the initial condition $\left|\begin{smallmatrix}1\\0\end{smallmatrix}\right|$ first, and then using the initial condition $\left|\begin{smallmatrix}1\\0\end{smallmatrix}\right|$. In the first case, we obtain the elements of the first column of the matrix, and with the second initial condition, the elements of the second column. Having calculated the elements of the Φ matrix, the continuity conditions of the tangential electromagnetic field components at the interfaces of separation of different materials have to be imposed, obtaining a homogeneous numerical system. The characteristic equation is derived by nullifying the Δ of the coefficients of the unknown parameters in the homogeneous system. The power normalization technique allows the calculation of the expression of all remaining unknowns.

2.3.2 *Channel waveguides*

In channel dielectric waveguides, light is confined in the transverse directions x and y, as shown in Fig. 2.2.

In this waveguide, only hybrid modes can propagate due to the difficulty of satisfying the boundary conditions at the interfaces, even if the longitudinal components of the electromagnetic field are less than the transverse ones. In particular, following the approximate modal technique [11], it has been demonstrated that the propagation of the modes E_{nm}^x, mainly polarized in the x-direction, and the modes E_{nm}^y mainly polarized in the y-direction, with m and n being the number of zeros in the field mapping.

The principle of operation of the channel waveguides is basically the same as slab waveguides, but the mathematical electromagnetic model is more complicated.

Different numerical techniques have been proposed to carry out the electromagnetic analysis of such guiding structures.

A very common numerical technique is the effective index method (EIM) [12], which is particularly useful to analyze non-homogeneous and

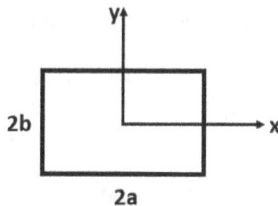

Fig. 2.2.   Channel dielectric waveguide.

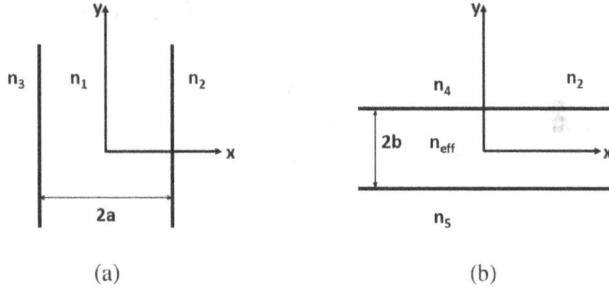

Fig. 2.3.   Effective index method applied to a channel waveguide. (a) Infinite slab wave-guide in y-direction; (b) infinite slab in x direction. $n_1 > n_2, n_3, n_4, n_5$; n_{eff} is the effective index of the guide (a).

anisotropic waveguides (e.g. Ti:LiNbO$_3$) [13] in combination with the transformation matrix technique, as described below.

When the EIM is used, the waveguide in Fig. 2.2 can be considered to be formed by two orthogonal slab waveguides, as shown in Fig. 2.3.

For the slab in Fig. 2.3(a), the characteristic equation of the TM modes corresponds to that of the E_{nm}^x modes of the channel waveguide:

$$\tan(2k_x a) = \frac{n_1^2 k_x \left(\varsigma_2 n_3^2 + \varsigma_3 n_2^2\right)}{\left(n_2^2 n_3^2 k_x^2 - n_1^4 \varsigma_2 \varsigma_3\right)}, \qquad (2.49)$$

where $\varsigma_i^2 = (n_1^2 - n_i^2)k_0^2 - k_x^2$ with k_0 the wave number in free space, k_x the x component of the wave number, and $i = 2, 3$.

Solving the problem with respect to $\mathbf{E}_x$, it is possible to define an effective index n_{eff} given by:

$$n_{\text{eff}}^2 = n_1^2 - (k_x / k_0)^2 \qquad (2.50)$$

This effective index is used to be the core refractive index for the guide in Fig. 2.3(b). For this last guide, the eigenvalue equation for TE modes is:

$$\tan(2k_y b) = \frac{k_y \left(\varsigma_{4e}^2 + \varsigma_{5e}^2\right)}{\left(k_y^2 - \varsigma_{4e} \varsigma_{5e}\right)}, \qquad (2.51)$$

where $\varsigma_{ie} = (n_{\text{eff}}^2 - n_i^2)k_0^2 - k_y^2$, with k_0 the wave number in free space, k_y is the y component of the wave number, and $i = 4, 5$.

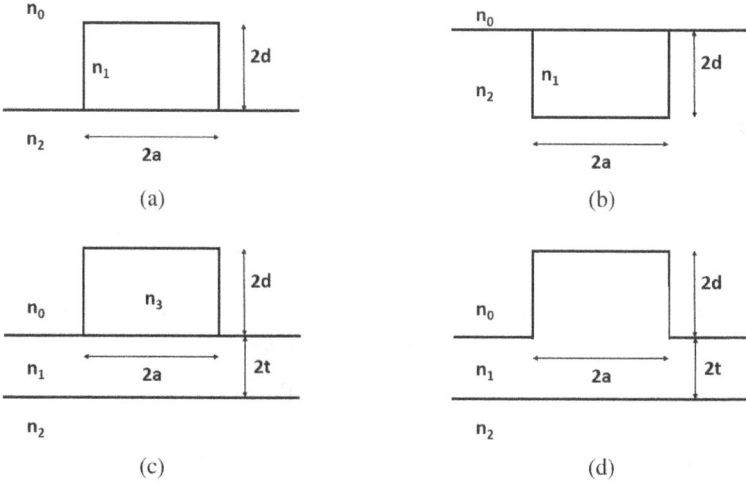

Fig. 2.4.   Cross-sections of 2D slab-coupled optical waveguides. (a) Raised strip guide; (b) embedded strip guide; (c) strip-loaded guide; (d) rib guide.

The propagation constant β is given by:

$$\beta^2 = k_0^2 n_{\text{eff}}^2 - k_y^2 \tag{2.52}$$

The same procedure is followed for the E_{nm}^y modes which characteristic equation of TE modes for the guide in Fig. 2.3(b) corresponds to:

$$\tan(2k_y b) = \frac{n_{\text{eff}}^2 k_y \left(\zeta_{4e} n_5^2 + \zeta_{5e} n_4^2\right)}{\left(n_4^2 n_5^2 k_y^2 - n_{\text{eff}}^4 \zeta_{4e}\zeta_{5e}\right)}, \tag{2.53}$$

where $\zeta_{ie} = (n_{\text{eff}}^2 - n_i^2)k_0^2 - k_y^2$ and $i = 4, 5$. The propagation constant can be determined as in (2.52).

Some common 2D optical waveguide configurations, which can be analyzed using the effective index method, are shown in Fig. 2.4.

2.3.3 *Optical fibers*

The first theoretical study on field propagation in dielectric cylindrical structures was carried out in 1910 [14], while the first experimental demonstration was given in 1966 [15]. The measured attenuation was around 1,000 dB/km, which has been reduced to fractions of dB/km in a few decades, following the development and remarkable success obtained through appropriate manufacturing processes, materials, and wavelengths.

2.3.3.1 *Step index optical fibers*

A cylindrical step index optical fiber is formed by a homogeneous core with a refractive index n_1 and radius a, embedded in a cladding of refractive index n_2 slightly lower than n_1, as shown in Fig. 2.5.

Assuming the wave propagation in the z-direction and the time operator to be $\exp(j\omega t)$, Maxwell's equations in cylindrical coordinates provide the following system for the core [16]:

$$\frac{1}{r}\frac{\partial H_z}{\partial \theta} - j\beta H_\theta = -jn_1^2\varepsilon_0\omega\, E_r,$$

$$j\beta H_r - \frac{\partial H_z}{\partial r} = -jn_1^2\varepsilon_0\omega\, E_\theta,$$

$$\frac{1}{r}\frac{\partial(rH_\theta)}{\partial r} - \frac{1}{r}\frac{\partial H_r}{\partial \theta} = -jn_1^2\varepsilon_0\omega\, E_z, \qquad (2.54)$$

$$\frac{1}{r}\frac{\partial E_z}{\partial \theta} - j\beta E_\theta = -j\mu_0\omega\, H_r,$$

$$j\beta E_r - \frac{\partial E_z}{\partial r} = j\mu_0\omega\, H_\theta,$$

$$\frac{1}{r}\frac{\partial(rE_\theta)}{\partial r} - \frac{1}{r}\frac{\partial E_r}{\partial \theta} = j\mu_0\omega\, H_z.$$

Of course, the same system where n_1 is substituted by n_2 is valid in the outer cladding. In Eq. (2.54), we express the transverse electromagnetic components as a function of the longitudinal ones E_z, H_z. Setting $d^2 = a^2(k_0^2n_1^2 - \beta^2)$ and $g^2 = a^2(\beta^2 - k_0^2n_2^2)$, the following wave equations can be derived for the core and cladding:

CORE $\qquad \dfrac{\partial^2 T}{\partial r^2} + \dfrac{1}{r}\dfrac{\partial T}{\partial r} + \left(\dfrac{d^2}{a^2} - \dfrac{v^2}{r^2}\right)T = 0 \qquad$ for $r \le a$, $\qquad (2.55)$

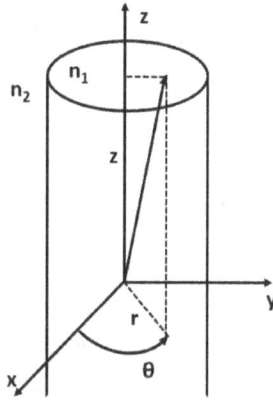

Fig. 2.5.   Cylindrical step index optical fiber.

CLADDING $\dfrac{\partial^2 T}{\partial r^2} + \dfrac{1}{r}\dfrac{\partial T}{\partial r} - \left(\dfrac{g^2}{a^2} - \dfrac{v^2}{r^2}\right)T = 0$ for $r > a$,

where $T(r)$ is defined by $\Psi(r, \theta) = T(r)\exp(jv\theta)$ with $\Psi(r, \theta) = E_z, H_z$, and $v = 1, 2, \ldots$, represents the azimuth periodicity.

The solutions of system (2.54) are Bessel functions of the first kind $J_v(d)$ in the core, and Hankel functions $K_v(g)$ in the cladding. Once we have determined the expressions of E_z and H_z in the core and cladding, the electromagnetic transverse components can be found $E_r, E_\theta, H_r, H_\theta$.

By imposing the continuity conditions of the tangential components $E_z, H_z, E_\theta, H_\theta$ at the interface $r = a$, the eigenvalue equation is derived:

$$[J'_v(d)/(dJ_v(d)) + K'_v(g)/(gK_v(g))][J'_v(d)/(dJ_v(d)) + (1 - 2\Delta)K'_v(g)/(gK_v(g))]$$
$$= (v\beta/k_0 n_1)^2(V/dg)^4, \tag{2.56}$$

where J'_v and K'_v are the derivatives of the Bessel and the Hankel functions, and V is given by $V^2 = k^2 a^2 n_1^2 \dfrac{n_1^2 - n_2^2}{n_1^2}$, with k wave number.

If $v = 0$, one can obtain the characteristic equation for both TM and TE modes:

$$J_1(d)/(dJ_0(d)) + (1 - 2\Delta)K_1(g)/(gK_0(g)) = 0 \quad \text{for TM}, \tag{2.57}$$

$$J_1(d)/(dJ_0(d)) + K_1(g)/(gK_0(g)) = 0 \quad \text{for TE}. \tag{2.58}$$

This means that for $\nu = 0$ the fiber supports TE_{0m} and TM_{0m} modes, while for $\nu \geq 1$, the fiber only supports hybrid modes:

— for $\nu > 0$, Helical (Skew) Modes (EH) propagate in the optical fiber with characteristic equation

$$J_{\nu+1}(d)/(dJ_\nu(d)) + K_{\nu+1}(g)/(gK_\nu(g)) = 0 \qquad (2.59)$$

— for $\nu > 0$, we find Helical (Skew) Modes (HE) modes, whose characteristic equation is:

$$J_{\nu-1}(d)/(dJ_\nu(d)) + K_{\nu-1}(g)/(gK_\nu(g)) = 0. \qquad (2.60)$$

Cut-off conditions can be easily derived for $\nu = 0$, $\nu > 1$ and $\nu = 1$. In this last case, the HE_{1m} modes are determined and, in particular, the HE_{11} mode, which is the fundamental mode for homogeneous cylindrical step index optical fibers.

2.3.3.2 Graded index optical fibers

The following index profile is assumed

$$n^2(r) = \begin{cases} n_1^2 \left[1 - 2\Delta f(r)\right] & \text{for } r < a, \\ n_1^2 \left[1 - 2\Delta\right] \equiv n_2^2 & \text{for } r < a, \end{cases} \qquad (2.61)$$

where n_1 is the value of the refractive index for $r = 0$, n_2 is the cladding index, and $2\Delta = \dfrac{n_1^2 - n_2^2}{n_1^2}$, and the function $f(r)$ may assume a generic mathematical expression. Most considered profiles are those with:

$$f(r) = (r/a)^a, \qquad (2.62)$$

where a is a numerical parameter whose value can be used for minimizing the differences of the intermodal delay in a multimodal fiber.

No analytical solutions for the electromagnetic field or propagation constants are available for any profile following the distribution law (2.62). Therefore, approximated numerical techniques or more complicated vectorial theory under weak guiding conditions, are used to extract the solutions. However, we have demonstrated [17] that the transformation matrix technique, already described in Section 2.3.1, can also be successfully applied to an optical fiber having any index profile, providing very accurate results in an ultra-short CPU time.

2.4 Optical resonators

An optical resonator is a device able to confine and store the energy at resonance frequencies, without leaving the light to escape outside. Typical optical resonator configurations are illustrated in Fig. 2.6.

Based on the physical operating principle, an optical resonator can be used as an optical filter or a spectrum analyzer in application fields such as sensing and telecoms, even in Space.

In this section, we refer to the simplest optical resonator configuration consisting of two parallel planar mirrors from which the light is constantly reflected, i.e. the so-called Fabry–Perot interferometer that is the key element of a laser source, and to ring resonators, which are the core of a number of very advanced devices in space applications.

To describe the whole operation principle of an optical resonator, we follow wave optics, which is useful to determine the characteristics of propagating modes at resonance. When spherical mirrors are used in a resonator, beam optics should be developed. Fourier optics can be utilized to study the diffraction effect at the mirrors, ray optics could also represent a good way to determine the appropriate geometrical parameters to confine the light in a resonator.

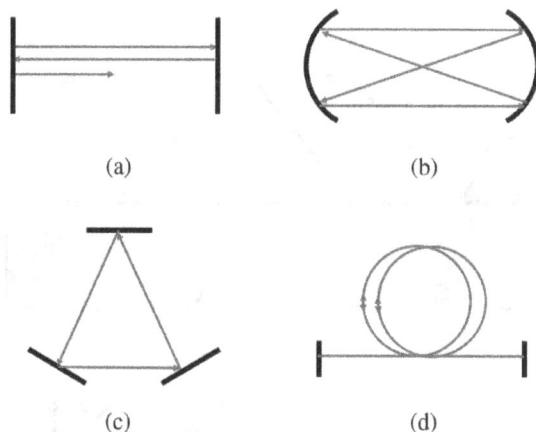

(a)   (b)

(c)   (d)

Fig. 2.6.   Optical resonators: (a) planar-mirror resonator, (b) spherical-mirror resonator, (c) ring resonator, (d) optical-fiber resonator.

2.4.1 *Multiple beam interferometry*

The approach described in Refs. [18–20] is useful to introduce multiple beam interferometry. This technique can produce an interference pattern generated by multiple coherent beams using two parallel partially reflecting mirrors. When light impinges on one of the two optical interface, it is repeatedly partially reflected and partially transmitted, so that multiple reflections can be observed together with an infinite number of transmitted and reflected rays. All transmitted rays are parallel and differ by a constant phase [21].

Figure 2.7 shows a thin optical layer having a refractive index n embedded in a medium with refractive index n'; E_0 is the electric vector complex amplitude of the incident plane wave, r is the reflection coefficient outside the optical layer, t is the transmission coefficient in the layer, r' is the reflection coefficient inside the layer, t' is the transmission coefficient from the layer to the upper medium, d is the distance between the mirrors, θ is the incidence angle and θ' is the transmission angle inside the thin layer.

It can be demonstrated that the phase difference due to the optical path between two adjacent rays is given by

$$\varphi = \frac{4\pi}{\lambda_0} nd \cos\theta', \qquad (2.63)$$

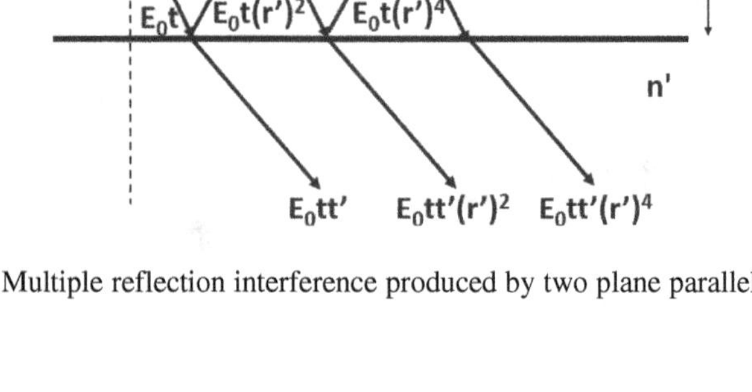

Fig. 2.7.   Multiple reflection interference produced by two plane parallel mirrors.

where λ_0 is the vacuum wavelength. However, since after each reflection an additional phase contribution φ_r [19] must be considered, the total phase difference is $\varphi_t = (\varphi + \varphi_r)$.

Assuming all transmitted contributions are summed, we obtain:

$$I_t = I_0 T + I_0 T R \exp(j\varphi_t) + I_0 T R^2 \exp(j2\varphi_t) + I_0 T R^3 \exp(j3\varphi_t) + \cdots, \quad (2.64)$$

where $r = -r'$, $r^2 = r'^2 = R$, $tt' = T = 1 - R$ and I_0 and I_t are the intensity at the input and the intensity transmitted, respectively. The total transmitted intensity can be derived from this equation after simple manipulation:

$$I_t / I_0 = T^2 / [(1-R)^2 + 4R\sin^2\varphi_t / 2] \quad (2.65)$$

which can be obtained by focusing the resulting transmitted light with a lens on a screen.

If no absorption is assumed (energy conservation), then

$$T + R = 1$$

and maximum and minimum transmission values can be derived:

$$I_{MAX} = T^2 / (1 - R)^2, \quad (2.66)$$

$$I_{MIN} = T^2 / (1 + R)^2. \quad (2.67)$$

The total reflected intensity is obtained by calculating the total reflected amplitude by summing all reflected waves:

$$E_r = E_0 r [1 - \exp(j\varphi_t)] / [1 - R\exp(j\varphi_t)], \quad (2.68)$$

which after some manipulations becomes:

$$I_r / I_0 = 4R\sin^2(\varphi_t / 2) / [(1-R^2) + 4R\sin^2(\varphi_t / 2)]. \quad (2.69)$$

The frequency spacing of adjacent peaks is equal to $f_s = c/2d$, with $c = c_0/n$ the speed of the light in the optical layer with refractive index n. At resonance, the length of the interferometer is an integer multiple of $\lambda/2$, i.e. $d = m\lambda/2$.

2.4.1.1 *Key parameters*

The quality factor for the reflectivity is defined as

$$Q_R = 4R/(1 - R)^2. \quad (2.70)$$

Obviously, at the peaks of the transmitted function, no reflection can be observed even with highly reflecting surfaces.

The contrast factor C is defined as the ratio between the intensity peak and intensity minimum:

$$C = I_{MAX}/I_{MIN} = [(1 + R)/(1 + R)]^2 = 1 + F. \tag{2.71}$$

The finesse F is a numerical value used to characterize the sharpness of the maxima. It is the ratio of the distance between peaks to full width at half maximum (FWHM).

The free spectral range (FSR) is the difference in wavelength between two successive maxima of the transmission function. It can be proved that in a filter

$$\text{FSR} = \lambda_0^2/2nd\cos\theta'. \tag{2.72}$$

The relation between FSR and FWHM is

$$\text{FSR} = F \times \text{FWHM}, \tag{2.73}$$

with $\text{FWHM} = f_r/F$ when $F \gg 1$.

2.5 Optical microresonators

The remarkable rise in materials science and technological processes (e.g. lithography) has allowed the growth in research efforts in the development of optical devices and systems. Among the large variety of optical devices now manufacturable, there are also good optical quality dielectric microcavities, which have already proved their great potential in a wide range of applications [22], including passive and active integrated optical devices [23–25], logic gates [26], and sensors [27].

In this section, we describe waveguiding integrated ring resonators for different applications, such as optical filters, lasers, sensors, and modulators. Among the different advantages of ring resonators, it should also be considered that they do not require end facets or gratings and are ready to be monolithically integrated with other components on a single chip.

The standard configuration of a channel dropping filter [28] based on an integrated ring resonator is shown in Fig. 2.8. Two bus waveguides can

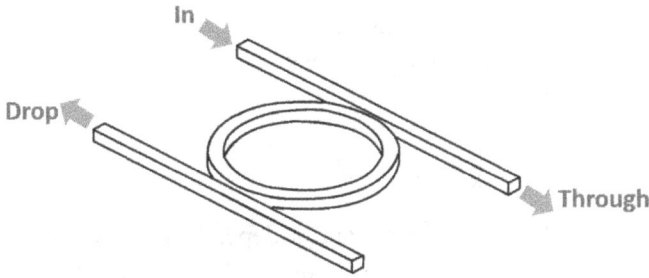

Fig. 2.8.   Ring resonator channel dropping filter.

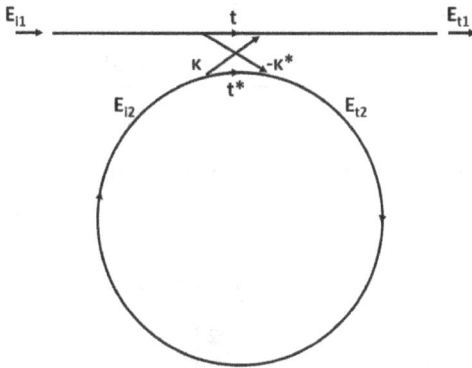

Fig. 2.9.   Ring resonator coupled to a single waveguide.

be coupled to the ring either through the evanescent electromagnetic field (directional couplers) or using multimode interference couplers.

2.5.1 *Single bus ring resonators*

A simpler and more efficient configuration for some applications is formed using just one bus straight waveguide (notch filter) [29] (see Fig. 2.9).

Usually, any looped cavity with a circular shape is denoted as a ring resonator, while the term racetrack resonator refers to a loop with a straight section in the coupler. In this book, we discuss ring resonators even though most physical aspects apply also to racetracks and loops with any other shape.

Under the conditions of single mode waveguides, lossless coupling, and normalized field amplitude, one can write [25]

$$\begin{pmatrix} E_{t1} \\ E_{t2} \end{pmatrix} = \begin{pmatrix} t & \kappa \\ -\kappa^* & t^* \end{pmatrix} \begin{pmatrix} E_{i1} \\ E_{i2} \end{pmatrix}, \tag{2.74}$$

where the coupling parameters t and κ depend on the coupling mechanism, and $\kappa_2 = -\kappa_1^*$ (* denotes the conjugated complex value). Moreover, $|k^2| + |t^2| = 1$ because the device is reciprocal.

After some mathematical manipulations, assuming $E_{i1} = 1$, it is possible to find the expressions of E_{i2}, E_{t1}, and E_{t2}, and thus, the power transmitted at the output of the bus:

$$P_{t1} = |E_{t1}|^2 = \frac{\alpha^2 |t|^2 - 2\alpha |t| \cos(\theta + \varphi_c)}{1 + \alpha^2 |t|^2 - 2\alpha |t| \cos(\theta + \varphi_c)}, \tag{2.75}$$

where a is the loss coefficient of the ring (zero loss: $\alpha = 1$), $\theta = \omega L / c_0$, $L = 2\pi r$, and $t = |t| \exp(j\varphi_c)$, with $|t|$ and φ_c that represent the coupling losses and the phase, respectively.

The power circulating in the ring is given by:

$$P_{i2} = |E_{i2}|^2 = \frac{\alpha^2 \left(1 - |t|^2\right)}{1 + \alpha^2 |t|^2 - 2\alpha |t| \cos(\theta + \varphi_c)}. \tag{2.76}$$

At the resonance condition: $(\theta + \varphi_c) = 2\pi m$, P_{t1} and P_{i2} become

$$P_{t1} = |E_{t1}|^2 = \frac{\left(\alpha - |t|\right)^2}{\left(1 - \alpha |t|\right)^2}, \tag{2.77}$$

$$P_{i2} = |E_{i2}|^2 = \frac{\alpha^2 \left(1 - |t|^2\right)}{\left(1 - \alpha |t|\right)^2}. \tag{2.78}$$

The critical coupling condition at which the transmitted power vanishes (destructive interference) is obtained when $\alpha = |t|$ in Eq. (2.77), i.e. when the propagation loss in the ring is equal to the loss in the coupler.

A typical transmission spectrum of the device in Fig. 2.9 is shown in Fig. 2.10.

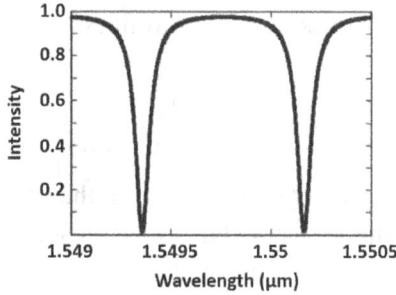

Fig. 2.10.   Notch type ring resonator spectrum.

2.5.2 Key parameters

Some key parameters for the single ring cavity can be defined as follows. The FSR in this case is given by:

$$\text{FSR} = -\frac{2\pi}{L}\left(\frac{\partial \beta}{\partial \lambda}\right)^{-1} \approx \frac{\lambda^2}{n_{\text{eff}}L}, \tag{2.79}$$

with $\beta = kn_{\text{eff}} = 2\pi n_{\text{eff}}/\lambda$, and the effective index of the waveguide is assumed constant with the wavelength.

If the effective index varies with the wavelength, the group index n_g has to be considered:

$$n_g = n_{\text{eff}} - \lambda \frac{\partial n_{\text{eff}}}{\partial \lambda} \tag{2.80}$$

and it is

$$\text{FSR} = \frac{\lambda^2}{n_g L}. \tag{2.81}$$

Another important parameter is the full width at the half maximum of the resonance peak, FWHM, which is also called 3-dB bandwidth of the resonance peak. It is given by

$$\text{FWHM} = \frac{\kappa \lambda^2}{\pi n_{\text{eff}}L}. \tag{2.82}$$

The finesse is the following ratio:

$$F = \frac{\text{FSR}}{\text{FWHM}}. \tag{2.83}$$

The quality factor Q is closely related to finesse and to the sharpness of the resonance peaks:

$$Q = \frac{\lambda}{\text{FWHM}} = \frac{\pi n_{\text{eff}} L}{\lambda} \frac{t}{1-t^2}. \tag{2.84}$$

Q is related also to the energy in the ring, in the sense that it is the ratio between the energy in the ring and the power lost in each optical cycle.

The intensity enhancement factor I_e

$$I_e = \left(\frac{E_{i2}}{E_{i1}} \right)^2 \tag{2.85}$$

is a measure of how much higher the energy in the ring is than in the bus waveguide, due to the constructive interference at resonance between the input wave and the guided mode in the ring. It should also be noticed that the field also experiences a phase shift of 2π in each round trip.

As already mentioned, the resonance condition is a wavelength-sensitive effect, which is particularly useful in sensing applications. So, the sensitivity parameter can be defined as the resonance wavelength shift due to a variation of any physical or geometrical parameter of the structure under analysis, which in turn causes a change in the effective index of the propagating mode in the device.

2.5.3 *Double bus ring resonators*

The configuration in Fig. 2.8 can include either a ring or a disk cavity between the two parallel dielectric straight waveguides. This configuration is known as an add-drop filter. The straight waveguides form four ports to connect the device to external components.

The operation of the device can be described by the interaction of harmonic optical waves propagating along the straight waveguide and the cavity, and by the resonance effect of the waves in the ring.

When a single mode is launched at the input port of the resonator, it propagates along the straight waveguide to the right, connecting the input port and throughput port, and part of it is evanescently coupled to the ring. While propagating in the ring, part of this signal is coupled to the straight waveguide to the left, and appears at the drop port. The remaining part of the signal propagates along the cavity, and interferes with the new input signal in the right coupling region, where these two fields may constructively or destructively interfere. The critical coupling condition occurs when $k_1 = k_2$ in a lossless device, while, when losses in the cavity are taken into account, it occurs when the losses are equal to the coupled energy.

The key parameters defined for the ring coupled to a single straight waveguide, i.e. FSR, FWHM, finesse F and Q factor, are also valid for this configuration. We would just remind you here the physical meaning of finesse and Q factor. They are related to the number of round trips made by the signal before spending its energy in losses and in the straight waveguides [30]. In particular, the finesse is the number of round trips made by the field in the cavity before its energy has been reduced to $1/e$ of the initial value, while Q is the number of oscillations of the field before the reduction of the energy in the cavity to $1/e$ of the initial value.

Since propagation, bend and coupling losses have to be reduced to obtain high Q factors, a ring resonator with a single straight waveguide has a higher Q factor value than a cavity with two bus waveguides, under the same operating conditions.

2.6 PhCs

In a crystalline material, which can be regarded as a periodic spatial arrangement of atoms or molecules, electrons show a periodic electrical potential that, following the principles of quantum mechanics, can be assumed as electromagnetic waves with a certain wavelength. Under specific conditions, the lattice can inhibit the propagation of a range of wavelengths in some directions, creating a photonic band gap (PBG) within the lattice band structure, analogously to electronic properties of semiconductor materials.

The most significant property of a PhC is its ability to manipulate the light on a wavelength or lesser scale. This property allows both theoretical and experimental demonstrations of interesting new functional devices, such as filters, waveguides, and resonant cavities, for several applications, e.g. sensing, telecom, biomedical.

The basic physics and main application fields of PhCs are described in this section.

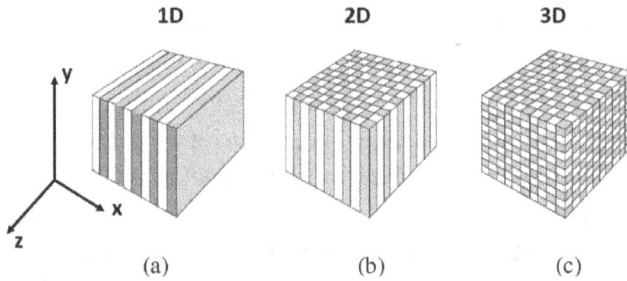

Fig. 2.11. PhC structures. (a) 1D, (b) 2D (hexagonal lattice of rods), (c) 3D (opal structure).

A lot of research effort has been spent from the late 1980s when the first papers were published [31, 32], and since then, a huge number of papers and textbooks have been published. 1D, 2D, and 3D PhCs, illustrated in Fig. 2.11 have been extensively investigated.

1D PhCs consist of alternating layers having different refractive index in which the PBG is formed only for specific incidence angles thus limiting its practical use. In a 2D PhC, a periodicity can be observed along two axes, while the device appears as homogeneous in the third direction. Here the PBG can be formed along in direction in the periodicity plane. A complete PBG can be obtained along in direction in 3D PhCs.

2.6.1 *Basic principles of PhCs*

In this section, we follow the mathematical developments in Ref. [33]. An electomagnetic analysis is required to describe the optical properties of the PhCs. Periodic dielectric media are characterized by periodic values of permittivity $\varepsilon(r)$ and permeability $\mu(r)$, which are also responsible for some changes in the characteristics of the optical wave with respect to the homogeneous medium. In particular, for example, the propagation constant is no longer proportional to the angular velocity and the propagating modes are Bloch waves instead of plane waves.

Maxwell's equations in a PhC are

$$\nabla \times E(r,t) = -\frac{1}{c}\frac{\partial B(r,t)}{\partial t}, \qquad (2.86)$$

$$\nabla \times H(r,t) = \frac{1}{c}\frac{\partial D(r,t)}{\partial t} + \frac{4\pi}{c}J(r,t), \qquad (2.87)$$

with B as the magnetic induction field $[\nabla \cdot B(r,t) = 0]$, D as the electric displacement $[\nabla \cdot [D(r,t)] = 4\pi\rho(r)]$ and $\rho(r,t)$ and $J(r,t)$ as the free electric charge and current densities, respectively.

Assuming harmonic electric and magnetic vector fields using the time operator $\exp(-j\omega t)$, a non-magnetic medium, and $\rho = J = 0$, Maxwell's equations can be simplified as:

$$\nabla \times E(r) - j\omega\mu_0 H(r) = 0, \qquad (2.88)$$

$$\nabla \times H(r,t) + j\omega\varepsilon_0\varepsilon(r)E(r,t) = 0. \qquad (2.89)$$

2.6.2 *Simulation algorithms*

From these last equations, one can derive the wave equation to be solved in the device. Solving the electromagnetic problem in 2D and 3D PhCs is a rather complicated procedure and, therefore, a number of modeling techniques have been proposed both in frequency- and time-domain methods. In frequency-domain methods, transmission spectra are calculated with reference to plane waves, and each field distribution is characterized by a single frequency, while in time-domain methods time iterations of Maxwell's equations are carried out.

Among the various algorithms utilized for analyzing the electromagnetic field in PhCs, we must very briefly mention the most common and well-known numerical techniques.

The plane wave method (PWM) [34] is based on the Fourier expansion of the fields and of the periodic dielectric function in terms of harmonic functions. The calculation of the fields is made only assuming the structure to be infinite in size.

The supercell method (ScM) [35] is applied to study periodic structures in slab waveguides by repeating the original PhC slab in the vertical direction.

The Bloch–Floquet method (BFM) [36] is useful to investigate 3D structures formed by a 2D periodicity layer and a homogeneous portion of material in orthogonal direction to the propagation axis. This method also allows calculation of leaky modes, modes at frequencies within the band gaps, and out-of-plane losses in very short computer time with high accuracy [37].

The finite difference time domain (FDTD) is a well-known method to analyze the propagation of electromagnetic waves in a generic structure. Some issues are represented by the quite large CPU time required for determining the characteristics of PhCs, mainly for identifying the band diagrams, when a good level of accuracy is required.

The finite element method (FEM), similar to FDTD, is a general method that can solve any kind of structure, requiring a very large computation time due to the very dense grids needed if high accuracy has to be achieved.

The transfer matrix method (TMM) [38] is an approximate method that assumes a number of layers to be invariant with respect to the propagation direction. Although this method is easy to apply, it is not a very efficient technique, because the computation time and the memory required grow rapidly with the longitudinal dimension of the device to be investigated.

The scattering matrix method (SMM) [39] cannot be applied to investigate large PhC devices because the computation effort scales with the third power of the holes (or rods) number in the device.

Green's functions method (GFM) [40] is a powerful technique for solving Maxwell's equations if the boundary conditions are known in the presence of a point source for exciting the structure. A new model of 2D PhCs based on Green's functions has been proposed in Ref. [41]. The analysis of the physical effects occurring in a multilayered structure including a periodic distribution of scatterers has been carried out in Ref. [42].

2.6.3 *Properties of PhCs*

The properties we refer to here are those that allow PhCs to be thought of as useful devices in several application fields. They are scaling, translation symmetry, and PBG.

Scaling

Scaling in frequency is a property of PhCs derived from the same property of the electromagnetic field. From Eq. (2.89), the wave equation

$$\nabla \times \left[\frac{1}{\varepsilon(r)} \nabla \times H(r) \right] = \left(\frac{\omega}{c} \right)^2 H(r),$$ (2.90)

can be easily obtained. If we introduce a scale change in frequency for the permittivity $\varepsilon(r)$, as $\varepsilon'(r') = \varepsilon(r/s)$, with $r' = sr$, $\nabla' = \nabla/s$, Eq. (2.90) becomes:

$$\nabla' \times \left[\frac{1}{\varepsilon'(r')} \nabla' \times H(r'/s) \right] = \left(\frac{\omega}{sc} \right)^2 H(r'/s)$$ (2.91)

which is the wave equation for $H(r'/s)$, with $\omega' = \omega/s$. Then we conclude that the solution of the wave equation can be scaled in size and wavelength.

Translational symmetry

Translational symmetry allows the crystal to maintain its properties invariant if it translates in a certain direction by a vector t. Under these conditions, for any t an operator x is defined so that

$$\xi\varepsilon(r) = \varepsilon(r + t) = \varepsilon(r).$$ (2.92)

When this property results continuous in all directions in the device, the solutions of the wave equation can be expressed as plane waves, all polarized in the same direction.

For the specific case of PhCs, it is possible to state that they are characterized by a discrete translational symmetry only along the periodicity directions. This implies that the system is invariant if the lattice vector $\mathbf{L}$ is an integer multiple of the lattice constant a in all directions:

$$\mathbf{L} = ma\mathbf{v}_1 + pa\mathbf{v}_2 + qa\mathbf{v}_3,$$ (2.93)

where $\mathbf{v}_1, \mathbf{v}_2, \mathbf{v}_3$ are the primitive vectors of the lattice. In this case, it is also possible to find an operator ξ_d so that:

$$\xi_d\varepsilon(r) = \varepsilon(r + t) = \varepsilon(r).$$ (2.94)

2.6.4 *Reciprocal lattice*

As already mentioned, due to translational symmetry, the plane wave only shows the same lattice periodicity at discrete wave vector **k** values. The periodicity can be expressed by:

$$\exp[j\mathbf{S}\cdot(\mathbf{r}+\mathbf{L})]=\exp(j\mathbf{S}\cdot\mathbf{r}),$$

from which it can be derived

$$\exp(j\mathbf{S}\cdot\mathbf{L})=1, \tag{2.95}$$

where **S** is the vector satisfying the periodicity condition. From Eq. (2.95), one also obtains

$$\mathbf{S}\cdot\mathbf{L}=2\pi l,$$

with l as the integer number.

It can be demonstrated that two waves interfere if

$$\mathbf{L}\cdot(\mathbf{k}-\mathbf{k}')=2\pi l,$$

where **k** is the wave vector of the incident wave and **k**′ is the wave vector of the diffracted wave, respectively. From this last equation, we have **S** = (**k** − **k**′), which represents the set of reciprocal lattice vectors **R**. All the reciprocal vectors are in the Brillouin zone, which is the elementary cell of the reciprocal lattice. The generic vector of the reciprocal lattice is:

$$\mathbf{R}=maw_1+paw_2+qaw_3,$$

where w_1, w_2, w_3 are the basic vectors of **R**.

2.6.5 *Photonic band gap*

The PBG configuration depends on the relation between the wave vector k and frequency ω. While in a homogeneous material system the diagram (ω, k) is a line with slope proportional to the refractive index, in a periodic medium, it can show one or more band gaps.

After simple manipulations, one can obtain:

$$(jk+\nabla)\times\left[\frac{1}{\varepsilon(r)}(jk+\nabla)\times\mathbf{u}_k(r)\right]=\left[\frac{\omega(k)}{c}\right]^2\mathbf{u}_k(r), \tag{2.96}$$

where $H_k(r) = e^{j(k \cdot r)} u_k(r)$, with $u_k(r)$ a function having the same periodicity as the lattice. By imposing $u_k(r) = u_k(r + L)$, the eigenvalues of Eq. (2.96) suffer a discretization and there will exist a discrete number of solutions at the frequency $\omega_i(k)$. The function $\omega_i(k)$ determines the PBG configuration.

The physical mechanism of formation of the PBG can be explained considering, as an example, a 1D PhC, consisting of a periodic stack of quarter-wavelength dielectric layers having alternating values of the refractive index. This grating is also called Bragg grating (see section 2.6.6). Under Bragg conditions, when the light impinges in a direction normal to the layers plane, all reflections from each layer contribute constructively to form the reflected wave, perfectly equal to the incident wave in a lossless structure. Out of the Bragg conditions a transmission wave can be observed because of the not fully constructive effect of reflected contributions. Since the Bragg condition depends on the wavelength it is clear that a PBG can be realized. The band gap also depends on the index contrast and it can be tailored as required by the specific application.

In a 1D PhC, following the physical aspect of the variational principle of the electromagnetic field, it can be observed that the energy carried by the modes at low frequency is more concentrated in regions with a higher refractive index, while the energy of modes at higher frequency is confined in regions with a lower refractive index. This is the reason for which in the transmission spectra there are lower dielectric bands, where the refractive index is high, and upper air bands at higher frequency, as in Fig. 2.12.

In a 2D PhC structure, a PBG can be formed for any propagation direction of the light in the periodicity plane. Under these conditions, due to the properties of the PhCs, decoupling of the two light polarizations occurs, which means that different transmission spectra and different band gaps can be derived for the two polarizations.

In a 2D PhC, when a wave propagates in it, depending on the physical and geometrical parameters of the configuration under investigation, some physical phenomena can be observed, such as reflection, transmission, diffraction and interference, and this may surely justify the formation of a band gap.

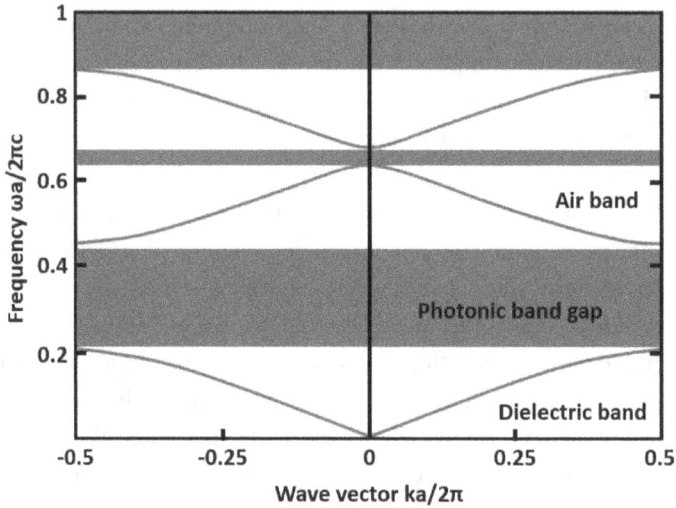

Fig. 2.12.   Band diagram of a 1D PhC formed by air–polystyrene multilayer.

In a 3D PhC, band gaps can be observed for any propagation direction of light, mainly if the structure shows high refractive index contrast and for certain lattice geometry (e.g. face-centered cubic lattice).

2.6.6 Defects in PhCs

A periodicity defect in a PhC gives rise to the possibility of having a number of modes at wavelengths within the band gap. The defect is simply obtained by adding or removing an amount of material or substituting a portion of material with a different refractive index, in the structure. Of course, intrinsic properties of periodic structures are invalid and the definition of an electromagnetic model has to be pointed out to investigate the effects of the defect.

Point defects are discontinuities well localized in the periodic structure. They can be created in 1D, 2D, and 3D PhCs, producing, for example, high quality factor and small mode volume cavities. Extended defects are also possible by removing one or several lattice rows in a fixed direction, as in the formation of a waveguide in a 2D PhC.

2.6.7 *Applications of PhCs*

A PhC showing a clear band gap, after creating in it an appropriate defect, could be used to realize optical functional devices, such as mirrors [43], waveguides [44], microcavities [45], or laser sources [46].

If the light incident on a PhC with no defects has a wavelength in the photonic gap, it is fully reflected. High efficiency mirrors can be manufactured with reflectivity >90%, optimizing the structure in order to limit the losses.

A PhC with a line defect realizes a waveguide supporting propagating modes at wavelengths in the band gap. Waveguides with sharp bends and very low losses have been both theoretically and experimentally demonstrated.

Compact and efficient PhC microcavities can be realized by modifying one or more holes (or rods) in a 2D PhC. In this way, modes at wavelengths within the band gap will be confined in the defect region.

The quality factor Q of such microcavities is given by

$$Q = \frac{2\pi W_s}{W_1} \tag{2.97}$$

with W_s as the stored energy and W_1 as the energy lost. Q may also be expressed in terms of wavelengths

$$Q = \frac{\lambda_0}{\Delta\lambda}, \tag{2.98}$$

where λ_0 is the center wavelength and $\Delta\lambda$ is the FWHM of the transmission peak.

Due to the physical properties of PhC microcavities, a strong field confinement together with a low loss is achievable. This characteristic has been exploited to demonstrate the feasibility of very compact laser sources with narrow pulse width and very good spectral purity because of the possibility of controlling the amplified spontaneous emission.

2.7 Passive waveguide components

Two broad categories of waveguiding components can be identified: passive components, where the characteristics of optical waves are investigated, and functional or active devices that are able to control the optical waves using energy of a different nature (e.g. electrical energy).

In this section, some passive guiding components are described. In particular, among the numerous components proposed in the literature in the last few decades, we will consider branching waveguides, directional couplers and diffraction gratings, which represent the basic building blocks for any integrated optical circuit.

2.7.1 *Branching waveguides*

A single mode branching waveguide is shown in Fig. 2.13. This optical component is also named Y-branch and is used to split or combine the power launched into a straight input waveguide.

The refractive index of the input waveguide and branch waveguide 1 and 2 is n_1 and n_2, respectively, and E_i, E_1, and E_2 are the electric fields of the guided modes in the input and two branches waveguides. The tapered section is needed to avoid the excitation of the lateral mode. To mitigate the scattering effect at the branching zone, the slanted angle θ cannot become so large that the field overlap between the input waveguide and two branches would not be reduced. For example, for a single mode Ti:LiNbO$_3$ waveguide, with $w = 4$ μm, $\Delta n = 5 \times 10^{-3}$ and a depth of 2 μm at $\lambda = 0.6328$ μm, the output power vanishes with $2\theta = 7°$ [47]. Depending on the value of the branching angle, the Y junction can operate as a mode splitter or power divider.

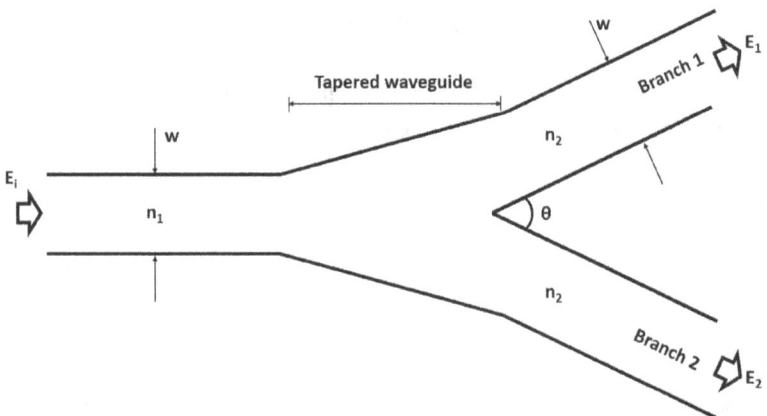

Fig. 2.13.   Sketch of a Y-branch waveguide.

2.7.2 *Directional couplers*

In an optical directional coupler including two identical single mode waveguides close to each other with a spacing of the order of a number of wavelengths, the optical power incident in one waveguide can be fully transferred to the output of the second waveguide, as illustrated in Fig. 2.14.

β_1 and β_2 are the propagation constants of waveguides 1 and 2, respectively, and L is the length of the coupling region, where odd and even modes are supported, after being excited by the incident wave in waveguide 1. After the optical path with length L, odd and even modes show a phase shift of π, therefore at the output of the coupling region ($z = L$), the field distribution becomes equal to that in waveguide 2, i.e. the total power is transferred from waveguides 1 to 2. The coupling coefficient depends on the propagation constants of odd and even modes, and the power transfer is related to the difference between the propagation constants of the two waveguides β_1 and β_2.

2.7.3 *Bragg gratings*

A Bragg grating is a key element in many devices for optical communications and sensing applications. Despite its well-known usefulness, the

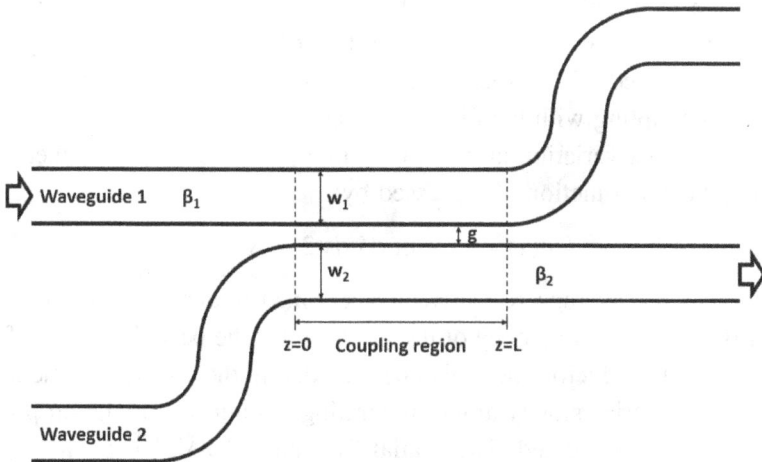

Fig. 2.14.   Directional coupler.

device is rather simple. In fact, it has a 1D PhC structure and interacts with light through the mode coupling mechanism. Some functions of Bragg gratings can be achieved by optimizing their parameters, e.g. period, modulation depth. Most Bragg gratings are known as fiber Bragg gratings (FBG), which are gratings realized on an optical fiber to be used for different application fields. In this subsection FBGs will be described.

The first photo-induced FBG was proposed in 1978 [48], and since then a huge research effort both theoretically and experimentally has been employed.

An FBG consists of a periodic modulation of the refractive index in the core of a single mode optical fiber, with the phase fronts perpendicular to the fiber axis. This modulation is most commonly achieved by varying the refractive index or the physical dimension of the fiber core. At each change of the refractive index, reflection of the propagating wave occurs giving rise to multiple reflections due to index modulation. The period of index modulation at the wavelength of the propagating light affects the relative phase of all partially reflected waves. Under the Bragg condition, occurring at a specific wavelength, the contributions of reflected light from each grating plane add constructively in the backward direction to form a back reflected peak with center wavelength depending on the grating period. Reflected contributions at other wavelengths do not add constructively and can be transmitted by the grating.

The period and amplitude of the grating can be controlled along the whole length of the grating in order to tailor its spectrum. Grating planes can also be tilted with respect to the propagation direction, which generates light coupling with the fiber cladding.

If the index variation in the grating in Fig. 2.15 can be described by a cosine, the $n(x)$ function is expressed by:

$$n(x) = n + \Delta n(x) \cos(2\pi x/\Lambda), \qquad (2.99)$$

where n is the average refractive index, $\Delta n(x)$ is the modulation depth, $2\pi/\Lambda$ is the spatial frequency of the grating, Λ is the period or pitch of the grating. A phase factor could also be included in the cosine argument.

If $\Delta n(x)$ varies slowly along the grating, an index modulation profile or apodization is created. The spatial frequency $2\pi/\Lambda$ changes along the

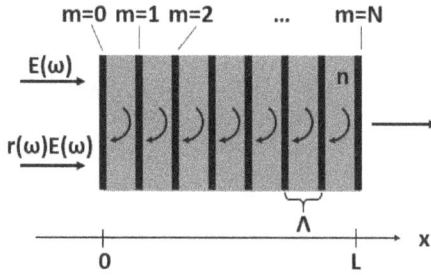

Fig. 2.15. Grating of length L and period Λ. Reflected wave depends on the reflection coefficient $r(\lambda)$.

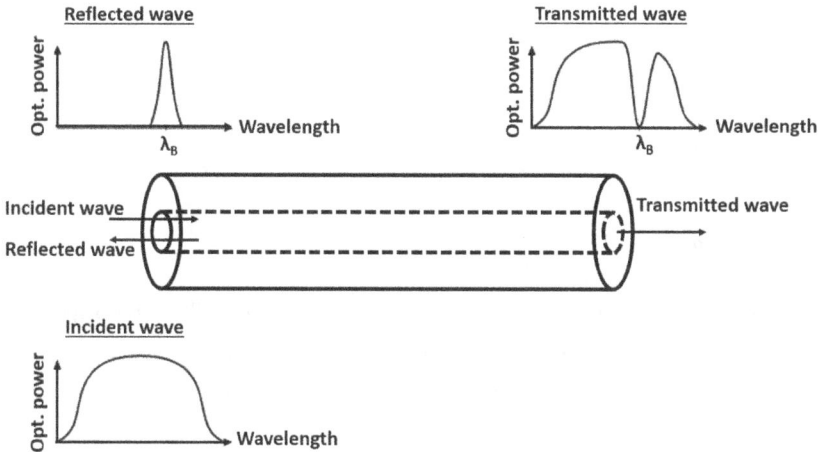

Fig. 2.16. Operating principle of a Bragg grating.

grating in chirped gratings. When the incident light has a wavelength satisfying the Bragg condition

$$\lambda_B = 2n_{eff}\Lambda, \qquad (2.100)$$

it will have the same phase as the reflected wave. n_{eff} is the effective refractive index of the fiber.

Figure 2.16 illustrates the property of the Bragg gratings.

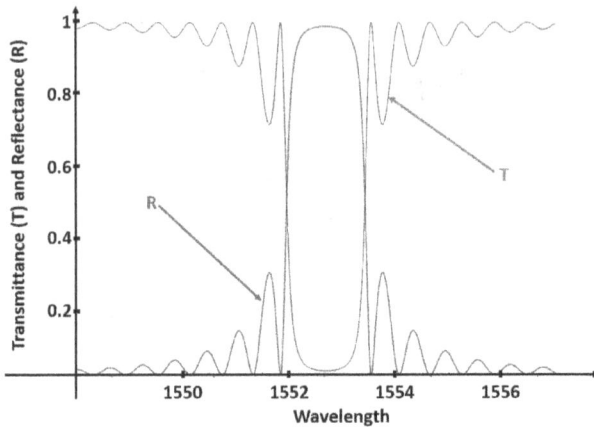

Fig. 2.17.   Reflection and transmission spectra of an FBG.

The reflection and transmission spectra of an FBG are sketched in Fig. 2.17.

The characteristics of Bragg gratings, which basically are reflectance filters, include easy manufacturing, low insertion loss, high wavelength selectivity, polarization insensitivity and full compatibility with single mode optical fibers. Filtering bandwidth as low as 0.1 nm or higher, e.g. 10 nm or more, can be achieved. The application fields of Bragg gratings are fiber grating sensors, fiber lasers, fiber optic communications. Several applications of Bragg gratings are reported in Ref. [49].

A remarkable research effort has been focused on theoretical investigation on Bragg grating characteristics to define the dependence of reflectivity on the wavelength. Two methods are commonly used to analyze and design these devices: the matrix transfer function method (TFM) and the coupled mode theory (CMT). In the TFM, the grating is assumed to be a stratified medium formed by layers with alternating refractive index value, under the general assumption of weakly guiding condition. Transverse modes in this case can be observed and a closed solution can be achieved. The CMT considers the grating as a perturbation in terms of refractive index, which couples the power of incident and reflected modes. Comparisons between TFM and CMT have been reported in order to highlight the differences [50, 51].

2.8 Electro-optic waveguide modulators

An electro-optic modulator is a device in which a strong interaction between the optical wave and the modulating electrical signal occurs.

In this section, electro-optical modulators in both ferroelectric (LiNbO$_3$) and semiconductor (InGaAsP/InP) materials will be described.

2.8.1 *Lithium niobate modulators*

Since the early 1980s, lithium niobate has been recognized as an excellent electro-optic material to produce high optical quality single mode waveguides through Ti indiffusion [8, 9, 52–56]. For this reason, all the waveguiding optical devices based on a ferroelectric material are manufactured utilizing a Ti indiffused LiNbO$_3$ waveguide [57].

Since then, a great number of functional LiNbO$_3$ devices have been proposed. Among them, attention was devoted to electro-optic modulators because of their good performance in terms of low drive voltage and broadband modulation. LiNbO$_3$ waveguide modulators are still used as external devices in DWDM communication systems to avoid frequency chirping as indirect modulation sources.

2.8.1.1 *Electro-optic effect in LiNbO$_3$*

When an electrical field is applied to the LiNbO$_3$ crystal, some physical effects can arise, such as a change of the refractive indexes of the material, electro absorption, photo refractivity, and nonlinear optical effects. In the following, only the change of refractive index is considered, and, in particular, the analysis will be limited to the Pockels' effect.

If a field E is applied, the change of the refractive index is given by

$$n(E) = n + a_1 E + \frac{1}{2} a_2 E^2 + \cdots, \qquad (2.101)$$

with

$$n = n(0), \ a_1 = \frac{dn}{dE}\bigg|_{E=0}, \ a_2 = \frac{d^2 n}{dE^2}\bigg|_{E=0} \cdots \qquad (2.102)$$

By setting $r = -2a_1/n^3$ and $s = -a_2/n^3$, we derive $a_1 = -\frac{rn^3}{2}, a_2 = -sn^3$, and then

$$n(E) = n - \frac{1}{2} rn^3 E - \frac{1}{2} sn^3 E^2 + \cdots. \qquad (2.103)$$

Since the terms with a power of $E > 1$ are $\ll n$, it is possible to neglect them.

With reference to the electrical impermeability $\eta = \varepsilon_0/\varepsilon = 1/n^2$, it is possible to write:

$$\Delta\eta = \frac{d\eta}{dn}\Delta n = -\frac{2}{n^3}\left(-\frac{1}{2}rn^3 E - \frac{1}{2}sn^3 E^2\right) = rE + sE^2, \quad (2.104)$$

and therefore

$$\eta(E) = \eta + rE + sE^2, \quad (2.105)$$

with $\eta = \eta(0)$, and r and s as the electro-optical coefficients whose values depend on the direction of the applied electrical field and light polarization. If the second order electro-optical effect is neglected, we get

$$\Delta(1/n^2)_{ij} = r_{ijk} E_k, \quad (2.106)$$

where r_{ijk} are the elements of the linear electro-optic 3rd rank tensor of the material, and E_k is the electrical field applied along the k direction.

When the propagation direction is normal to the optic axis of the LiNbO$_3$ crystal, the index change is:

$$\Delta n' = -(n'^3/2)r'E', \quad (2.107)$$

where n' is a combination of the refractive indexes, r' is a linear sum of the electro-optic coefficients and E' is an appropriate component of the electric field. Consequently, when an electrical field is applied to a medium, the phase shift induced on a wave at the wavelength λ, propagating over a distance L in the crystal is given by:

$$\Delta\varphi = \Delta\beta \cdot L = \Delta n \, 2\pi L/\lambda \quad (2.108)$$

with β the propagation constant and n the refractive index.

The half-wave voltage, i.e. the voltage necessary to obtain a phase shift of π over a length L of the material, is expressed as:

$$V_\pi = \frac{\lambda}{n^3 r}. \quad (2.109)$$

2.8.1.2 *Modulation characteristics*

In order to modulate an optical signal, a controlled change of the optical characteristics of the material (e.g. LiNbO$_3$) needs to be generated where the wave propagates. This means that the size of the index ellipsoid, which describes the spatial index distribution, would change and this can be done by applying an external electrical or mechanical field. To this end, it is possible to exploit the electro-optic, acousto-optic, and magneto-optic properties of lithium niobate to design and manufacture modulators.

In electro-optical modulators, it is possible to modulate the phase (or frequency), polarization, or intensity.

Phase modulation is realized in most modulators, but it requires a phase coherent detection system [58–62]. To overcome this issue, polarization modulation can be utilized, where a simple polarization sensitive detector or an analyzer can be placed ahead of the detector.

Intensity modulation can also be derived from phase modulation of two orthogonal optical signals [58].

The main operating parameters of modulators are the modulation depth, bandwidth, power consumption, isolation, and insertion loss.

Modulation depth, or modulation index, for intensity modulators, is given by

$$m_d = (I_0 - I)/I_0 \text{ if } I_0 > I, \tag{2.110}$$

where I is the transmitted light intensity with the electrical signal applied, and I_0 is the intensity without the electrical signal [60].

The extinction ratio, or maximum modulation depth, is:

$$(m_d)_{max} = (I_M - I_0)/I_M \text{ if } I_M \geq I_0, \tag{2.111}$$

where I_M is the transmitted intensity with the maximum value of the electrical signal. Modulation depth for phase modulators can be also derived. For interferometric modulators, the modulation depth is equal to [60]:

$$m_d = \sin^2(\delta\varphi/2), \tag{2.112}$$

with $\delta\varphi$ as the phase change.

The bandwidth B of a modulator, by convention, is the frequency range between the two up and down frequency where m_d is reduced to 50% of its maximum value, and of course it represents the frequency range over which the device can operate with high efficiency.

As for the power consumption, an important figure of merit given by P/B [mW/MHz] needs to be evaluated, where P is the drive power. Of course, the design of an electro-optic modulator must be carried out by minimizing the power consumption.

The isolation parameter in a modulator is related to the maximum value of the modulation depth. It is a measure of the isolation between devices at the input and output of the modulator.

The insertion loss L_i can be calculated as:

$$L_i = 10 \log (I_w/I_0), \tag{2.113}$$

where I_w is the intensity transmitted in the waveguide without the modulator.

2.8.1.3 *Bulk modulators*

Electro-optical modulators in $LiNbO_3$ can be classified as bulk or wave-guiding modulators. In most bulk modulators based on the electro-optic effect, a spontaneous polarization or phase modulation occurs.

The simplest configuration of a bulk modulator is shown in Fig. 2.18.

In a homogeneous $LiNbO_3$ crystal with electrodes deposited on two parallel surfaces, the light propagating in the crystal suffers from the diffraction effect due to the size of the square cross-section. If L is the length and d the thickness (i.e. the gap between the electrodes) of the crystal, the following expression can be derived:

$$\frac{d^2}{L} = S^2 \frac{4\lambda}{n\pi}, \tag{2.114}$$

where S is a numerical coefficient with value ranging from 3 to 6, related to the attenuation of light while propagating through the crystal.

Assuming a vertical optical axis of the lithium niobate, wave propagation in direction perpendicular to the z-axis and the electric field direction as in Fig. 2.18, it is possible to exploit the electro-optical effect at its maximum value if the wave is polarized in the optical axis direction.

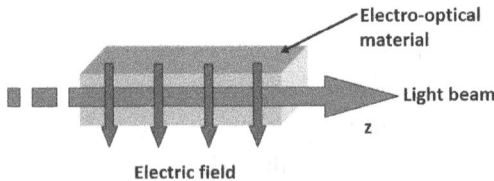

Fig. 2.18.   Simple configuration of a bulk modulator.

2.8.1.4 *Waveguide modulators*

Waveguide modulators (also known as thin film modulators) allow a strong light confinement together with a strong interaction between optical and electrical fields in the waveguide, thus increasing the device performance.

Waveguide modulators in different electro-optic materials have been proposed, and the best performance has been demonstrated for $LiNbO_3$ modulators based on a Ti-indiffused waveguide.

Device configurations with a rectangular waveguide have shown very good modulation efficiency and large bandwidth.

The criteria to be followed in the design of thin film $LiNbO_3$ modulators are:

— Manufacturing of a very good optical quality waveguide in lithium niobate;
— Choice of the most appropriate crystal orientation and the direction of light propagation, in order to get the maximum value of the electrooptic effect;
— Choice of the electrode configuration depending on the functionality required in the device;
— Highly efficient coupling from the laser source to the modulator and from the modulator to the external optical fiber;
— Low power consumption.

A comparison between bulk and waveguide modulators is reported in Ref. [60].

2.9 Instabilities in $LiNbO_3$ optical devices

$LiNbO_3$ single-crystals, in the ferroelectric state, show a spontaneous polarization along the +c optic axis which is reversible under certain conditions. Investigations on ferroelectric microdomains have been carried out [63, 64], also for estimating the long-term DC drift in lithium niobate optical devices [65].

A large number of devices have been used in telecommunications, optical signal processing, and sensing systems. A number of commercial devices are also available.

Total instability in lithium niobate waveguide devices is defined as the effect of some physical phenomena such as the photorefractive effect, thermal drift, as well as both short- and long-term DC drift.

The sources of instability and possible solutions mainly for electro-optical modulators, which must to be stable in the operating conditions and environment for use in optical communication systems, are analyzed in this subsection.

2.9.1 *Photorefractive effect*

Optical damage in lithium niobate waveguides, i.e. the power degradation and unstable characteristics, is caused by the photorefractive effect. It has been widely accepted that the optically induced refractive index changes in lithium niobate can be attributed to some impurity dopants and, particularly, to iron, whose density defines the absorption which has its maximum at $\lambda = 0.45$ μm. Fe^{2+} ions act as traps ($Fe^{2+} + h\nu \leftrightarrow Fe^{3+} + e^-$) creating intermediate electronic energy states in the crystal band gap. When photons in the visible range are injected, electrons are generated and their transition to the conduction band can be observed along the positive optical axis c because of the photovoltaic effect. Thus, a charge separation is produced, which creates a refractive index change due to the electro-optic effect. Moreover, the refractive index change causes the effective index and, the electromagnetic field distribution to change.

In Ref. [66], theoretical and experimental results are presented on the damage thresholds in Ti-indiffused Z-cut $LiNbO_3$ waveguides.

The photorefractive effect changes the electromagnetic field configuration in a $Ti:LiNbO_3$ channel waveguide. For a waveguide that is 3 μm wide, the optical damage appears at input density powers of 20 W/cm² at a visible wavelength (~0.5 μm).

An important concern for lithium niobate devices to be employed in long haul optical fiber communications systems is, of course, their resistance to optical damage. Optically induced index change is more severe at relatively low wavelengths. In fact, it has been observed that optically induced drift occurs with input powers as low as 3.2 μW at 0.850 μm. At 1.3 μm, a threshold power of 0.5 mW has been reported. In Ref. [67], a study of the stability of $LiNbO_3$ Mach–Zehnder (MZ) modulators at 1.51 μm is reported for launch power levels less than 75 mW. Insertion loss, optical extinction, phase change and switching voltage have been measured, for increasing optical power over a 7 h period of measurement. They had no evidence of significant changes in those parameters, thus confirming that complex feedback control electronics is not required to compensate for the drift effect.

Lithium niobate bulk crystals with congruent composition show a rather low optical damage resistance (~0.3 GW/cm^2) compared with other optical materials. Therefore, lithium niobate congruent crystals should not be used in applications involving high power light pulses. Two possible ways of increasing the damage threshold are ~5% MgO doping or the use of pure crystals with stoichiometric composition [68].

The photorefractive effect implies photon absorption at appropriate wavelengths, which, in turn, generates some thermally induced refractive index changes. However, the index change Δn_t induced by the thermal effect is proportional to the temperature change with a constant ratio $\partial n/\partial T = -0.7 \times 10^{-4}/°C$. It has been verified that Δn_t is negligible with respect to the index change due to the photorefractive effect, when typical values (≈ 10 μW) of the power are used.

2.9.2 Thermal drift

Thermal drift is the variation of the driving point due to temperature. Electro-optic guided wave devices in lithium niobate are required to exhibit large bandwidth and velocity matching of the RF and optical signals together with thermal stability. A thick buffer layer and a shielding plane on the electrodes are usually enough to achieve those requirements.

To eliminate the thermal instability, dual oxidized silicon films to be used as buffer layer have been proposed in Ref. [69] for a 20 GHz low driving voltage (<5 V) 2 × 2 Z-cut Ti:LiNbO$_3$ switch at 1.53 μm. In fact, when a single dielectric buffer layer (SiO$_2$) is used, the DC voltage induced drift can be suppressed [70]. In Ref. [71] a model of thermal instability in X-cut LiNbO$_3$ MZ modulators, which take into account the pyroelectric effect has been proposed at 1.55 μm. They have found that even small differences, Δw, in the width of the two waveguides induce the shift of the driving point, and the thermal instability due to the modification of the pyroelectric properties of the substrate material produced by the fabrication process, is proportional to Δw. They demonstrated that the increase in the surface conductivity at the waveguide–air interface gives a good thermal stability (i.e. a phase shift less than 5° in the temperature range from −10°C to + 25°C).

2.9.3 DC drift

DC drift is the shift of the driving point when a voltage is applied. It produces a degradation of the modulation efficiency and extinction ratio in lithium niobate guided-wave modulators. Short-term DC drift is a phenomenon occurring immediately after the application of the voltage (from several seconds to several minutes) while long-term drift is relevant to a longer period.

As already mentioned, an oxidized buffer layer is an effective solution of the problem of short-term DC voltage drift in lithium niobate devices.

An accurate study of the unstable phenomena and their temperature dependence is reported in Ref. [72] for MZ Ti:LiNbO$_3$ optical modulators in the 1.5 μm band.

Short-term DC drift results experimentally in an exponential shift. The polarity, amplitude, and relaxation time of short-term DC drift depend on the configuration of the buffer layer. Dual oxidized layers are effective in reducing the relaxation time and the driving point shift. As for the temperature dependence of the relaxation time of some unstable phenomena such as short-term DC drift, temperature drift (i.e. the variation of the driving point due to that of the environment temperature), and long-term DC drift, it has been found that their activation energies are almost equal and around 0.7 eV.

Long-term drift is a linear shift of the driving point, it is proportional to time and depends on the electric field applied to the waveguide. Its relaxation time (hours) is much longer than that of short-term drift and temperature drift. Relaxation times of the last two phenomena can be considered to be almost equal ($\leq$1 h). However, the time constants are, in general, highly dependent on the sample under test. The long-term drift, which is observed as a gradual shift of the driving point when a DC voltage is applied to the device, is believed to be associated with the gradual change of the space charge around the waveguides.

2.10 Electro-absorption modulators

Electro-absorption modulators are electro-optic modulators even if they do not exploit the Pockels effect. They are based on the Franz–Keldysh [73] or quantum confined Stark effect (QCSE) [74] in semiconductor materials. Electro-absorption modulators are attractive when integrability, high extinction ratio, low drive voltage and high speed modulation are required.

The Franz–Keldysh effect in bulk semiconductors can be explained with reference to the energy band diagram, observing that the band edges bend in the presence of a strong electric field, which should not allow transition to the conduction band. However, since the electric field increases the states in that band, a reduction of the effective band gap occurs. As a result, the absorption edge moves towards lower energies with increasing electric field.

In multi-quantum well (MQW) structures, the QCSE occurs, i.e. the transition energy for electrons and holes suffers a reduction when the electric field is applied in the direction perpendicular to the wells, with a variation of the absorption strength. Thick quantum wells show high modulation efficiency, while in thin quantum wells, stronger absorption can be observed. A contribution to the absorption is given by the formation of electron–hole pair systems (excitons).

In Ref. [75], as an example, high-performance MQW electro-absorption modulators with a peak extinction efficiency of 35 dB/V at 1.55 μm, have been demonstrated. The authors also reported a deep physical insight by using 2D simulations of carrier transport, optical absorption, and optical waveguiding.

2.11 Acousto-optic waveguide devices

Similar to electro-optic modulators, acousto-optic modulator operation is based on the formation of an index grating generated by an acoustic wave within a material with good acousto-optic properties. However, ultrafast responses comparable with those of electro-optic devices cannot be achieved in acousto-optic devices, mainly due to the different speed of the relevant waves in solids.

In most acousto-optic devices, surface acoustic waves (SAWs) [76] are utilized, with their characteristic of having the possibility of varying the wavelength continuously over a significant range depending on the frequency of the electrical signal. This characteristic is exploited in a number of acousto-optic tunable devices. Modulators, switches, mode converters, filters, and deflectors have been proposed in the literature. In Ref. [48], several acousto-optic waveguiding devices are described.

2.11.1 *Acousto-optic effect and interaction*

The acousto-optic effect is the change in the refractive index due to a mechanical strain induced in the material by an acoustic wave.

The index change in a piezoelectric material (e.g. quartz, $LiNbO_3$) can be expressed by:

$$\Delta(1/n^2)_{ij} = p_{i,j,k,l}\, \mathbf{S}_{k,l}, \tag{2.115}$$

where $\mathbf{S}_{k,l}$ is the strain tensor, and $p_{i,j,k,l}$ are the components of the photoelastic tensor. Terms of higher order are, in general, neglected.

When a SAW propagates in an acousto-optic modulator, a diffraction grating is generated, which can work under two different regimes, depending on the length L of the modulator and the wavelength of the acoustic and optical wave, Λ and λ, respectively. In the Raman–Nath regime, it results in $L \ll \Lambda^2/\lambda$ and numerous diffraction orders arise. In this case, the optical wave impinges transversely to the acoustic wave and the interaction length between the two waves is short. Thus, the optical waves suffer a phase grating diffraction, which produces interference peaks. Bragg diffraction occurs for $L \ll \Lambda^2/\lambda$ and just one diffraction order can be observed. This situation occurs when the acoustic beam is wider. The grating formed by a surface acoustic wave generated by a transducer causes the Bragg diffraction of an optical beam and the intensity of the diffracted wave depends on the product of the incident light intensity and the power of the acoustic beam [77]. In Ref. [60], some interesting acousto-optic devices are described, such as Raman–Nath modulators, Bragg modulators, beam deflectors and switches, as well as frequency shifters.

A piezoelectric–mechanical model has been proposed [78] for investigating the acousto-optic interaction and determining the propagation characteristics of surface waves in a generalized multilayered structure. The model can include any stress and strain fields in terms of elastic, piezoelectric, and dielectric tensors of the involved materials. In addition, the electro-optic and photoelastic tensors, using the model, can define a relation between the strain and electrical fields and the spatial periodic oscillation of the permittivity of the grating.

2.12 Semiconductor lasers

In this section, the basic physics, operation, and fabrication of some types of semiconductor lasers are briefly described. For further in-depth investigation, the reader is referred to a number of excellent papers and books on the subject [79–83].

The semiconductor laser has been commercially available for years and is the most used of all lasers in a large variety of applications, e.g. disc players and optical communication systems. Advances in the technological process have led to the fabrication highly efficient devices, even on subwavelength scale, with many types of configurations.

The key application field of laser diodes is in optical telecommunications [84]. Other applications refer to optical radars [85], optical recording [86], optical signal processing [87], pump sources [88], sensor systems [89], and medical equipments [90].

2.12.1 *Basic operation*

The principle of operation of a laser can be considered as an extension of an optical amplifier. In fact, while in an optical amplifier radiation passes through the active medium once, in a laser light, it undergoes many reflections at the end mirrors of the resonant cavity. To reach the lasing operation, the light amplitude needs to grow at each round trip and all round trip contributions interfere constructively. Therefore, one can state that lasing in all types of lasers is related to the gain mechanism (wave generation created by electron–hole recombinations, see Fig. 2.19) in the material and the presence of a resonant cavity. Electron–hole recombination, which can

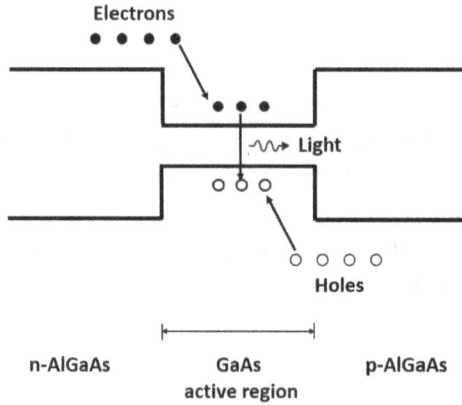

Fig. 2.19.   Recombination of electrons and holes in a semiconductor laser.

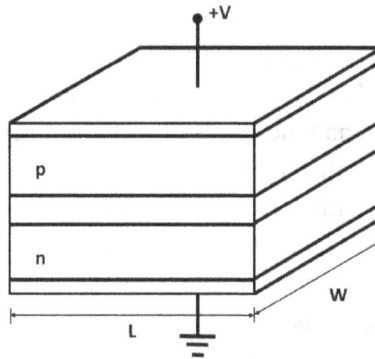

Fig. 2.20.   Basic structure of a laser diode.

be considered as the annihilation of a conduction electron and a valence hole, releases energy that is used to generate photons or transferred to the intrinsic structure of the material with consequent heating of the lattice.

The energy gap of the material is responsible for the wavelength of the generated light. Electrons and holes to be recombined are injected from *p* and *n* doped regions of a *p–n*-junction. Either optical or electric carrier pumping provides electrons and holes to be recombined. Direct band gap material is required to achieve efficient lasing operation.

A simple scheme of a *p–n* junction laser is illustrated in Fig. 2.20. Two parallel end faces act as mirrors to realize the required optical feedback.

Radiative recombination producing injection photoluminescence depends on spontaneous emission (band-to-band transitions). The presence of an electromagnetic wave at the right wavelength can produce stimulated emission. All photons emitted under stimulated emission conditions have the same wavelength and same phase as the stimulating wave, which means that the resulting emitted radiation is coherent with the stimulating light.

The absorption, spontaneous emission, and stimulated emission are schematically represented in Fig. 2.21.

Assuming E_1 and E_2 as energy levels in the valence and conduction band, respectively, it is possible to determine the Fermi–Dirac function both in the valence and conduction bands [91]:

$$f_v(E_1) = \frac{1}{e^{(E_1-F_v)/kT} + 1}, \qquad (2.116)$$

$$f_c(E_2) = \frac{1}{e^{(E_2-F_c)/kT} + 1}, \qquad (2.117)$$

where F_v and F_c are the quasi-Fermi levels in the valence and conduction band, respectively.

Starting from the previous two formulae and considering the electron and hole densities in the conduction and valence band, respectively, the increase of the energy density of the light and the contribution of the

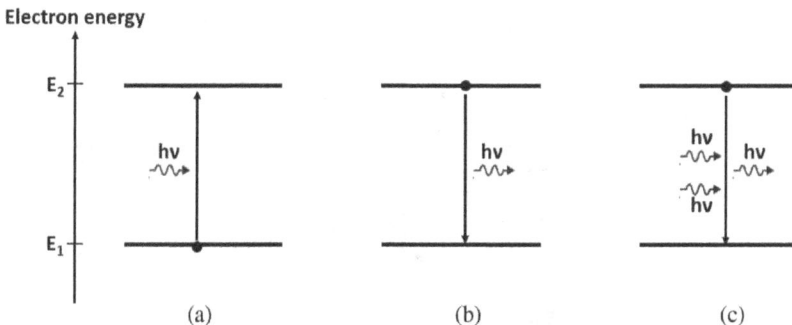

Fig. 2.21. (a) Absorption, (b) spontaneous emission, and (c) stimulated emission processes in a material system with only two electron energy levels.

spontaneous emission can be derived. Moreover, the gain expression is also derived:

$$g(h\nu) = B\frac{h\nu}{V}\int g_c(E_2)g_v(E_2 - h\nu)\left[f_c(E_2) - f_c(E_2 - h\nu)\right]dE_2, \quad (2.118)$$

where $h\nu$ is the photon energy, $g_c(E_2)$ and $(g_v(E_1))$ are the densities (per unit energy and volume) of the electron (hole) states in the conduction band at E_2 (E_1) level, and $E_1 = (E_2 - h\nu)$ [92].

The first semiconductor lasers [79–80] had a homojunction structure and were only demonstrated operating at low temperature. All lasers on the market include structures based on a double heterostructure (DHS) that conversely can operate at room temperature (see Fig. 2.22). The refractive index of the active region is always larger than the surrounding media to collect and confine the light. Thin active regions must be fabricated to improve the recombination process, while the surrounding media should have a much greater (three to four order of magnitude) thickness.

A brief historical review of the physics and technology of heterostructure lasers based on DHS and new heterostructure configurations is reported in Ref. [92].

End mirrors of the cavity can be formed by direct cleavage if a reflectivity value of around 30% is sufficient, or by depositing a thin film dielectric coating to achieve higher reflectivity values. The end reflecting faces, also named Fabry–Perot (FP) faces or surfaces, because they operate like FP mirrors, provide the necessary optical feedback that allows

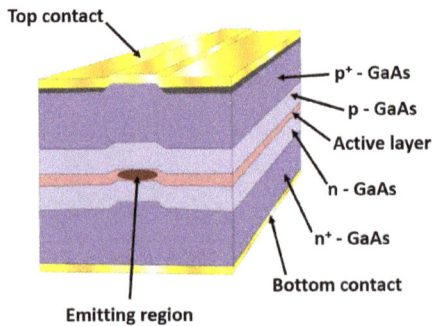

Fig. 2.22.   DHS laser operating in the continuous wave (CW) regime at room temperature [92].

one or more longitudinal modes to be established at resonance. When the laser diode is directly biased, the inversion of population occurs and light is generated by both stimulated and spontaneous emission processes. Generated photons travel back and forth many times perpendicularly to the FP surfaces to create a steady state operation to which one or more electromagnetic field distributions (modes) correspond. Thus, a standing wave is produced and there is a resulting integer number of half-wavelengths between the mirrors results. The mode number is equal to the number of half-wavelengths:

$$M = 2Ln/\lambda_0, \tag{2.119}$$

where the refractive index n is a function of λ and λ_0 the vacuum wavelength of the emitted light. Mode spacing can be evaluated by the derivative $dM/d\lambda_0$.

At each round trip in the cavity, the light intensity I, traveling along the direction x perpendicular to the mirrors, experiences an amplification. The light intensity I_1 can be evaluated using the following formula [91]:

$$dI_1/dx = gI_1 + \xi, \tag{2.120}$$

where $\xi = AN_2h\nu$. Here, A is Einstein's coefficient, N_2 is the number of carriers per unit volume at the energy level E_2, h is Planck's constant, ν is the frequency, and g is the gain.

Neglecting ξ, the power in the cavity is given by:

$$P = P_0 \exp(gx). \tag{2.121}$$

Moreover, the light amplitude amplification for one round trip is:

$$E = E_0 R\exp[1/2(g - \alpha)2L + j2\beta L, \tag{2.122}$$

with $R = r_1r_2$ (product of the reflectivity of the two mirrors), α is the attenuation constant and β is the propagation constant.

The amplitude condition under which the light can be amplified after each round trip is the following:

$$R\exp(g - \alpha)L \geq 1. \tag{2.123}$$

As for the phase condition, it is necessary that the phase of the light after one round trip be an integer multiple of 2π:

$$2\beta L = m2\pi, \ldots \quad m = 1,2,3,\ldots$$

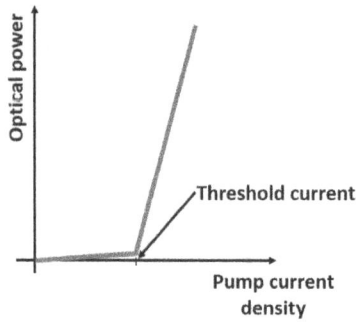

Fig. 2.23.   Optical characteristic of the laser diode.

It is impossible to realize a perfect single longitudinal mode laser. A number of longitudinal modes can be observed around the main peak. Modification of the laser structure can produce a spectrum formed by a very narrow peak at the desired wavelength.

Immediately after forward biasing the laser diode, no lasing effect can be observed because at low injection current levels only incoherent spontaneous emission is excited characterized by a relatively wide linewidth (hundreds of Angstroms). When the current level increases, stimulated emission is generated, which becomes dominating with respect to the spontaneous emission, and the linewidth can be very narrow (of the order of 1 Å). The light is now coherent. Thus, a threshold value of the current can be defined where the transition from incoherent to coherent emitted light occurs (see Fig. 2.23).

The threshold current value is obtained when the number of stimulated photons is equal to the overall number of photons lost. The procedure to derive the mathematical expression of the current density at the threshold, the output power and the overall power efficiency is derived in Ref. [60].

2.13 Semiconductor optical detectors

The best known type of semiconductor optical detector is the *p–n* junction diode. When it is reverse biased with a relatively large voltage, the reverse current has a contribution provided by the drift current in the depletion layer and a second contribution due to the diffusion current formed in the doped region of the junction. In the space charge region, the carriers (electrons and holes) are separated by the reverse electric field. Minority carriers can diffuse to the edges of the depletion region, while majority carriers remain in their (doped) regions.

For a junction with a well-defined voltage barrier, the calculated total current density J_{tot} is expressed [93] as the sum of the photocurrent and the leakage (or dark) current:

$$J_{tot} = e\Phi\left(1 - \frac{e^{-\alpha w}}{1 + \alpha L_d}\right) + e p_{n0} \frac{D_d}{L_d}, \qquad (2.124)$$

where e is the electron charge, Φ is the total photon flux, α the absorption coefficient, w the width of the depletion layer, L_d the diffusion length, D_d the hole diffusion constant, and p_{n0} the hole density at the equilibrium. The dark current does not depend on the photon flux. Photocurrent accounts for the drift and diffusion components without considering scattering loss and free-carrier absorption [60].

The quantum efficiency η is defined as the number of carriers generated for each incident photon:

$$\eta = 1 - \frac{e^{-\alpha w}}{1 + \alpha L_d}. \qquad (2.125)$$

Some improvements in the performance can be obtained if the depletion layer is embedded in a waveguide. In fact, the total current density becomes:

$$J_{tot}^w = e\Phi\,[1 - \exp(-\alpha L)], \qquad (2.126)$$

where L is the length of the photodiode. By tailoring the geometrical parameters and optimizing the voltage value, it is also possible to achieve a quantum efficiency of 100%. Moreover, since the capacitance can be very small, the high frequency response would be improved.

A number of different types of photodiodes are commonly used, such as Schottky barrier, avalanche, p–i–n and metal–semiconductor–metal (MSM) photodiodes [60].

A Schottky barrier photodiode is a conventional photodiode in which the depletion layer is replaced by a metal–semiconductor junction. The performance of this device is the same as that of a conventional photodiode and the above-mentioned expressions of the photocurrent density J_{tot} and J^w_{tot} are still valid.

In avalanche photodiodes, the value of the reverse bias voltage is very close to breakdown, which leads to a carrier multiplication caused by strong ionization, with a consequent high gain. These devices are also able to operate at very high frequency. However, particular attention is required in the technological process to fabricate avalanche photodiodes in terms of physical and geometrical parameters, which have to be designed in order to excite the avalanche effect.

The *p–i–n* photodiodes are characterized by a low carrier concentration in the depletion layer, which extends through the intrinsic layer to increase the product αw. A high-thickness intrinsic layer is also useful to enlarge the bandwidth.

MSM photodiodes show Schottky barriers formed on the surface of a semiconductor layer with an interdigitated configuration. Low capacitance values can be realized together with short transit times of carriers and, thus, large bandwidth.

References

[1] B. E. A. Saleh and M. C. Teich (2001). *Fundamentals of Photonics*. Wiley, New York.

[2] J.-M. Liu (2009). *Photonic Devices*. Cambridge University Press, Cambridge.

[3] D. Marcuse (1974). *Theory of Dielectric Optical Waveguides*. Academic Press, New York.

[4] M. J. Adams (1981). *An Introduction to Optical Waveguides*. Wiley, New York.

[5] H. Kogenlik (1979). Theory of Dielectric Waveguides, in *Integrated Optics*, T. Tamir (Ed.), Springer-Verlag, New York.

[6] M. N. Armenise, R. De Leo and M. De Sario (1976). Numerical analysis of propagating modes in inhomogeneous optical fibres. *Alta Frequenza*, vol. 45, pp. 230–234.

[7] M. N. Armenise and M. De Sario (1980). Investigation of the guided modes in anisotropic diffused slab waveguide with embedded metal layer. *Fiber and Integrated Optics*, vol. 3, pp. 197–219.

[8] M. N. Armenise (1989). Kinetics of Ti-indiffusion in lithium niobate, in *Properties of Lithium Niobate*. INSPEC, London, UK, pp. 253–256.

[9] M. N. Armenise (1989). Characteristics of lithium niobate waveguides, in *Properties of Lithium Niobate*. INSPEC, pp. 261–266.

[10] M. N. Armenise, C. Canali, M. De Sario, P. Franzoni, J. Singh, R. H. Hutchins and R. M. De La Rue (1984). Dependence of inplane scattering levels in $Ti:LiNbO_3$ optical waveguides on diffusion time, *IEE Proceedings H: Microwaves Optics and Antennas*, vol. 131, pp. 295–298.

[11] E. A. J. Marcatili (1969). Dielectric rectangular waveguide and directional coupler for integrated optics. *The Bell System Technical Journal*, vol. 48, pp. 2071–2102.

[12] R. M. Knox and P. P. Toulios (1970). *Integrated Circuits for the Millimetre Through Optical Frequency Range, Symposium on Submillimeter Waves*, New York (USA), March 31–April 2.

[13] M. N. Armenise and M. De Sario (1978). Modal characteristics of diffused rectangular anisotropic waveguides to design integrated optics devices, *8th European Microwave Conference*, Paris, September.

[14] D. Hondros, P. Debye (1910) "Elektromagnetische Wellen an dielektrischen Drähten", *Annalen der Physik*, vol. 337, pp. 465–476.

[15] C. K. Kao and G. Hockam (1966). Dielectric fiber surface waveguides for optical frequencies, *Proceedings of IEEE*, vol. 113, pp. 1151–1158.

[16] K. Iizuka (2002) *Elements of Photonics*, Chapter 11, Vol. 2. Wiley Interscience, New York.

[17] E. Born and M. Wolf (1999). *Principles of Optics: Electromagnetic Theory of Propagation, Interference and Diffraction of Light*. Cambridge University Press, Cambridge.

[18] G. R. Fowles (1990). *Introduction to Modern Optics*. Dover Publications Inc., New York.

[19] G. Hernandez (1988). *Fabry–Perot Interferometers*. Cambridge University Press, Cambridge.

[20] L. Giovannelli (2011). Study of a Fabry–Perot interferometer prototype for space applications, Master Thesis, University of Rome "Tor Vergata".

[21] K. J. Vahala (2003). Optical microcavities. *Nature*, vol. 424, pp. 839–846.

[22] E. A. J. Marcatili (1969). Bends in optical dielectric guides. *The Bell System Technical Journal*, vol. 48, pp. 2103–2132.

[23] F. C. Blom, D. R. van Dijk, H. J. W. M. Hoekstra, A. Driessen and T. J. A. Popma (1997). Experimental study of integrated optics microcavity resonators: Towards and all optical switching device, *Applied Physics Letters*, vol. 71, pp. 747–749.

[24] M. Kuwata-Gonokami, R. H. Jordan, A. Dodabalapur, H. E. Katz, M. L. Schilling, R. E. Slusher and S. Ozawa (1995). Polymer microdisk and microring lasers. *Optics Letters*, vol. 20, pp. 2093–2095.

[25] B. E. Little, S. T. Chu, H. A. Haus, J. Foresi and J.-P. Laine (1997). Microring resonator channel dropping filters, *IEEE Journal of Lightwave Technology*, vol. 15, pp. 998–1005.

[26] T. A. Ibrahim, R. Grover, L. C. Kuo, S. Kanakaraju, L. C. Calhoun and P. T. Ho (2003). All-optical AND/NAND logic gates using semiconductor microresonators. IEEE Photonics Technology Letters, vol. 15, pp. 1422–1424.

[27] R. W. Boyd and J. E. Heebner (2001). Sensitive disk resonator photonic biosensor, *Applied Optics*, vol. 40, pp. 5742–5747.

[28] A. Yariv (2002). Universal relations for coupling of optical power between microresonators and dielectric waveguides. *Electronics Letters*, vol. 36, pp. 321–322.

[29] D. G. Rabus (2007). *Integrated Ring Resonators The Compendium.* Springer, New York.

[30] W. Bogaerts, T. Van Vaerenbergh, K. De Vos, S. Kumar Selvaraja, T. Claes, P. Dumon, P. Bienstman, D. Van Thourhout and R. Baets (2012). Silicon microring resonators. *Laser & Photonics Reviews*, vol. 6, pp. 47–73.

[31] E. Yablonovitch (1987). Inhibited spontaneous emission in solid-state physics and electronics. *Physical Review Letters*, vol. 58, pp. 2059–2062.

[32] S. John (1987). Strong localization of photons in certain disordered dielectric superlattices. *Physical Review Letters*, vol. 58, pp. 2486–2487.

[33] C. Ciminelli (2011). Introduction to photonic crystals and metamaterials, in *Selected Topics on Metamaterials and Photonic Crystals*. World Scientific Publishing, pp. 1–46.

[34] M. Plihal and A. A. Maradudin (1991). Photonic band structure of two-dimensional systems: The triangular lattice. *Physical Review B*, vol. 44, pp. 8565–8571.

[35] F. Ramos-Mendieta and P. Halevi (1997). Electromagnetic surface modes of a dielectric superlattice: the supercell method. *Journal of the Optical Society of America B*, vol. 14, pp. 370–381.

[36] C. Ciminelli, F. Peluso and M. N. Armenise (2005). Modeling and design of two-dimensional guided-wave photonic band-gap devices. *IEEE Journal of Lightwave Technology*, vol. 23, pp. 886–901.

[37] C. Ciminelli, H. M. H. Chong, F. Peluso, M. N. Armenise and R. M. De La Rue (2004). High Q guided-wave photonic crystal extended microcavity, ECOC 2004, Post deadline paper. Stockholm, Sweden, September, pp. 26–27.

[38] J. B. Pendry (1994). Photonic band structure, *Journal of Modern Optics*, vol. 41, pp. 209–229.

[39] J. Yonekura, M. Ikeda and T. Baba (1999). Analysis of finite 2-D photonic crystals of columns and lightwave devices using the scattering matrix method. *IEEE Journal of Lightwave Technology*, vol. 17, pp. 1500–1508.

[40] D. N. Green and S. C. Bass (1984). Representing periodic waveforms with non-orthogonal basis functions, *IEEE Transactions on Circuits and Systems*, vol. 31, pp. 518–534.

[41] C. Ciminelli, R. Marani and M. N. Armenise (2009). Investigation of a point-like and plane-wave excitation in 2D photonic band gap microcavities using Green's function method, *Optical and Quantum Electronics*, vol. 41, pp. 255–265.

[42] C. Ciminelli, R. Marani and M. N. Armenise (2010). Fast and accurate investigation of 2-D multilayered photonic crystals by a 3-D model based on the Green's Function, *IEEE Journal of Quantum Electronics*, vol. 46, 1549–1560.

[43] D. Labilloy, H. Benisty, C. Weisbuch, T. F. Krauss, V. Bardinal and U. Oesterle (1997). Demonstration of cavity mode between two-dimensional photonic crystal mirrors, *Electronics Letters*, vol. 33, pp. 1978–1980.

[44] T. F. Krauss (2007). Slow light in photonic crystal waveguides. *Journal of Physics D: Applied Physics*, vol. 40, pp. 2666–2670.

[45] B. D'Urso, O. Painter, J. O'Brien, T. Tombrello, A. Yariv and A. Scherer (1998). Modal reflectivity in finite-depth two-dimensional photonic crystal micorcavities. *Journal of Optical Society of America B*, vol. 15, pp. 1155–1159.

[46] H. Altug, D. Emglund and J. Vukovic (2006). Ultrafast photonic crystal nanocavity laser. *Nature Physics*, vol. 2, pp. 484–488.

[47] H. Nishihara, H. Masamitsu and S. Toshiaka (1989). *Optical Integrated Circuits*. Mc Graw Hill, New York.

[48] K. O. Hill, Y. Fujii, D. C. Johnson and B. S. Kawasaki (1978). Photosensitivity in optical fiber waveguides: Applications to reflection filter gratings. *Applied Physics Letters*, vol. 32, pp. 647–649.

[49] C. R. Giles (1997). Lightwave applications of fiber Bragg gratings, *IEEE Journal of Lightwave Technology*, vol. 15, pp. 1391–1404.

[50] B.-G. Kim and E. Garmire (1992). Comparison between the matrix method and the coupled-wave method in the analysis of Bragg reflector structures. *Journal of Optical Society of America*, vol. 9, pp. 132–136.

[51] U. Bandelow and U. Leonhardt (1993). Light propagation in one-dimensional lossless dielectrica: Transfer matrix and coupled mode theory. *Optics Communication*, vol. 101, pp. 92–99.

[52] M. N. Armenise, C. Canali, M. De Sario, A. Carnera, P. Mazzoldi and G. Celotti (1982a). Evaluation of the Ti diffusion process during fabrication of Ti:LiNbO$_3$ optical waveguides. *Journal of Non-Crystalline Solids*, vol. 47, pp. 255–257.

[53] M. N. Armenise, C. Canali, M. De Sario, A. Carnera, P. Mazzoldi and G. Celotti (1982b). Ti-compound formation during Ti diffusion in LiNbO$_3$. *IEEE Transactions on Components, Hybrids, and Manufacturing Technology*, vol. 5, pp. 212–216.

[54] M. N. Armenise C. Canali, M. De Sario, A. Carnera, P. Mazzoldi and G. Celotti (1984). Ti diffusion process in LiNbO$_3$, in *New Directions in Guided Wave and Coherent Optics*, NATO A.S.I. Series, vol. 78/79, pp. 623–637.

[55] M. N. Armenise, C. Canali, M. De Sario, A. Carnera, P. Mazzoldi and G. Celotti (1983). Characterization of Ti$_{0.65}$Nb$_{0.65}$O$_2$ compound as a source for Ti diffusion during Ti:LiNbO$_3$ optical waveguides fabrication. *Journal of Applied Physics*, vol. 54, pp. 6223–6231.

[56] M. De Sario, M. N. Armenise, C. Canali, A. Carnera, P. Mazzoldi and G. Celotti (1985). TiO$_2$, LiNbO$_3$O$_8$, and (Ti$_x$Nb$_{1-x}$)O$_2$ compound kinetics during Ti:LiNbO$_3$ waveguide fabrication in presence of water vapours. *Journal of Applied Physics*, vol. 57, pp. 1482–1488.

[57] M. N. Armenise (1988). Fabrication techniques of lithium niobate waveguides. *IEEE Proceedings Journal of Optoelectronics*, vol. 135, pp. 85–91.

[58] I. P. Kaminow, L. W. Stulz and E. H. Turner (1975). Efficient strip-waveguide modulator. *Applied Physics Letters*, vol. 27, pp. 555–557.

[59] M. Kawabe, S. Hirata and S. Namba (1979). Ridge waveguides and electro-optical switches in LiNbO$_3$ fabricated by ion-bombardment-enhanced etching. *IEEE Transanctions on Circuits and Systems*, vol. 26, pp. 1109–1113.

[60] R. G. Hunsperger (2009). *Integrated Optics: Theory and Technology*. Springer, New York.

[61] M. N. Armenise and M. De Sario (1983). Traveling-wave electrooptical modulator in inhomogeneous lithium niobate. *Alta Frequenza*, vol. 52, pp. 212–214.

[62] D. Hall, A. Yariv and E. Garmire (1970). Observation of propagation cutoff and its control in thin optical waveguides, *Applied Physics Letters*, vol. 17, pp. 127–129.

[63] L. Huang and N. A. F. Jaeger (1994). Discussion of domain inversion LiNbO$_3$. *Applied Physics Letters*, vol. 65, pp. 1763–1765.

[64] S. Miyazawa (1979). Ferroelectric domain inversion in Ti-diffused LiNbO$_3$ optical waveguide, *Journal of Applied Physics*, vol. 50, pp. 4599–4603.

[65] T. Nozawa and S. Miyazawa (1996). Ferroelectric microdomains in Ti-diffused LiNbO$_3$ optical devices. JPN *Journal of Applied Physics*, vol. 35, pp. 107–113.

[66] J. C. Chon, W. Feng and A. R. Mickelson (1993). Photorefractive damage thresholds in Ti:LiNbO$_3$ channel waveguides. *Applied Optics*, vol. 32, pp. 7572–7580.

[67] A. R. Beaumont, C. G. Atkins and R. C. Booth (1986). Optically induced drift effects in lithium niobate electro-optic waveguide devices operating at a wavelength of 1.51μm. *Electronics Letters*, vol. 22, pp. 1260–1261.

[68] F. Abdi, M. Aillerie, P. Bourson, M. D. Fontana and K. Polgar (1998). Electrooptic properties in pure LiNbO$_3$ crystals from the congruent to the stoichiometric composition. *Journal of Applied Physics*, vol. 84, pp. 2251–2254.

[69] T. Nozawa, M. Yanagibashi, H. Miyazawa, K. Kawano and H. Jumonji (1990). A broadband, low driving voltage 2 × 2 Ti:LiNbO$_3$ optical switch, in *Photonic Switching II*, K. Tada and H. S. Hinton (Eds.), Springer-Verlag, Berlin, pp. 84–87.

[70] A. R. Beaumont, B. E. Daymond-John, W. A. Stallard and R. C. Booth (1986). A non-destructive technique for rapidly assessing the stability of lithium niobate electro-optic waveguide devices, topical meeting on integrated and guided-wave optics '86, Atlanta, GA, February 25–27, pp. 46–47.

[71] M. Varasi, M. Ricci, M. Signorazzi and A. Vannucci (1999). Thermal instability of LiNbO$_3$ modulators, *Proceedings of the 9th European Conference on Integrated Optics (ECIO '99)*, Turin, April 13–16, pp. 233–236.

[72] H. Jumonji and T. Nozawa (1992). Instabilities and their characterization in Mach-Zehnder Ti:LiNbO$_3$ optical modulators. *Electronics and Communications in Japan (Part II: Electronics)*, vol. 75, pp. 76–88.

[73] J. I. Pamkove (1971). *Optical Processes in Semiconductors*. Prentice Hall, New Jersey.

[74] D. A. B. Miller, D. S. Chemla, T. C. Damen, A. C. Gossard, W. Wiegmann, T. H. Wood and C. A. Burrus (1984). Band-edge electroabsorption in quantum well structures: The quantum-confined stark effect. *Physical Review Letters*, vol. 53, pp. 2173–2176.

[75] J. Piprek, Y.-J. Chiu, S.-Z. Zhang, J. E. Bowers, C. Prott and H. Hillmer (2002). High-efficiency MQW electroabsorption modulators, *Proceedings of the 1st International Symposium on Integrated Optoelectronics*, Philadelphia, Pennsylvania, USA, May 12–17, pp. 139–149.

[76] R. M. White (1970). Surface elastic waves. *Proceedings of the IEEE*, vol. 58, pp. 1238–1276.

[77] C. S. Tsai (1979). Guided-wave acousto-optic Bragg modulators for wide-band integrated optic communications and signal processing. *IEEE Transactions on Circuits and Systems*, vol. 26, pp. 1072–1098.

[78] M. N. Armenise, V. M. N. Passaro and F. Impagnatiello (1991). Acoustic-mode analysis of a homogeneous multilayer guiding structure. *Journal of Optical Society of America B*, vol. 8, pp. 443–448.

[79] R. N. Hall, G. E. Fenner, J. D. Kingsley, T. J. Soltys and R. O. Carlson (1962). Coherent light emission from gaAs junctions. *Physical Review Letters*, vol. 9, pp. 366–369.

[80] M. I. Nathan, W. P. Dumke, G. Burns, F. H. Dill and G. Lasher (1962). Stimulated emission of radiation from GaAs p–n junctions. *Applied Physics Letters*, vol. 1, pp. 62–64.

[81] H. C. Casey and M. B. Panish (1978). *Heterostructure Lasers, Part A: Fundamental Principle*. Academic Press, New York.

[82] G. H. B. Thompson (1980). *Physics of Semiconductor Laser Devices*. John Wiley & Sons, New Jersey.

[83] G. P. Agrawal and N. K. Dutta (1986). *Long–Wavelength Semiconductor Lasers*, Van Nostrand Reinhold, New York.

[84] S. E. Miller and A. G. Chynoweth (1988). *Optical Fiber Telecommunications*. Academic Press, 1988.

[85] G. L. Abbas, W. R. Babbitt, M. de La Chappelle, M. L. Fleshner, J. D. McClure and E. J. Vertatschitsch (1990). High-precision fiber-optic position sensing using diode laser radar techniques, *Proceedings of SPIE*, vol. 1219, pp. 468–479.

[86] R. A. Bartolini, A. E. Bell and F. W. Spong (1981). Diode laser optical recording using trilayer structures. *IEEE Journal of Quantum Electronics*, vol. QE-17, pp. 69–77.

[87] M. N. Armenise, E. Pansini and A. Fioretti (1990). Wave guide correlator system for real time radar data processing.

[88] R. L. Byer (1987). R. V. Pole Memorial Lecture: The renaissance in soild-state lasers, in *International Quantum Electronics Conference*, OSA Technical Digest, vol. 21, paper MI1.

[89] M. N. Armenise, V. M. N. Passaro, F. De Leonardis and M. Armenise (2001). Modeling and design of a novel miniaturized integrated optical sensor for gyroscope applications. *IEEE Journal of Lightwave Technology*, vol. 19, pp. 1476–1494.

[90] D. Conteduca, F. Dell'Olio, C. Ciminelli and M. N. Armenise (2015). New miniaturized exhaled nitric oxide sensor based on a high Q/V mid.infrared 1D photonic crystal cavity. *Applied Optics*, vol. 54, pp. 2208–2217.

[91] K. IIzuka (2002). *Elements of Photonics*, Chapter 14, Vol. 2. Wiley Interscience, New York.

[92] Z. Alferov (2000). Double heterostructure lasers: Early days and future perspectives. *IEEE Journal on Selected Topics in Quantum Electronics*, vol. 6, pp. 832–840.

[93] S. M. Sze (1981). *Physics of Semiconductor Devices*, 2nd Edn. Wiley, New York.

Chapter 3

Optical Links for Inter- and Intra-Spacecraft Communications

For more than three decades, optical fibers have been considered a valuable alternative to conductive wires to interconnect sub-systems on board a spacecraft (box-to-box interconnects) [1]. The optical communication links implemented using optical fibers provide a number of advantages over competing technologies, such as immunity to electromagnetic interference, transparency to any modulation/coding format, low signal distortion, extremely large bandwidth, low power consumption, ease of system integration/testing, less damaging of sub-systems during the integration activity, mechanical flexibility, low volume, and light weight. Since the electromagnetic energy is fully confined within the fiber core, optical fiber cables, which are formed by a number of fibers, do not interfere with each other, with wires or with other single fibers, thus notably simplifying spacecraft design, integration, and testing. Moreover, optical fibers offer the widest available bandwidth (several THz), the lowest signal attenuation, the smallest signal distortion, and are more affordable than all other wired physical media.

Due to these interesting advantages, at the beginning of the 1990s, the US National Astronautics and Space Administration (NASA) started to use fiber-optic data links for satellite telemetry and control applications. In the NASA Solar, Anomalous and Magnetospheric Particle Explorer (SAMPEX) satellite, which was launched on June 19, 1992, the control and data handling functions were performed by the Small Explorer Data System

(SEDS), whose telemetry and control bus was implemented using optical fiber technology [2]. Other NASA missions utilizing optical fiber data link technology include the Rossi X-ray Timing Explorer (RXTE), the Tropical Rainforest Measuring Mission (TRMM), and the Microwave Anisotropy Probe (MAP) [2]. This technology is also utilized on board the International Space Station, where an optical fiber data bus, called the High Rate Data Bus and operating at 0.1 Gb/s, was mounted [2]. In 2002, a European Space Agency (ESA) R&D program started to study photonics applications in the field of spacecraft engineering. During this program, communication links on board satellites were identified as one of the most promising applications of photonics for space [3].

Currently, optical links are used for data transfer or to distribute high-purity analog signals on board a spacecraft. In nearly all those links, the optical beam propagates in single/multi-mode optical fibers. A few optical wireless data links for intra-satellite communications have also been developed.

In the next years, data rates that space vehicles should support in two-way communications between them and from ground stations are expected to increase exponentially up to tens of Gb/s [4]. This trend suggests that radio frequency (RF)-based communications, which are currently the most reliable technology for space communications, will be partially replaced by the laser communication (lasercom) technology, which has already demonstrated its potential to face the challenge imposed by the growing data volume. Lasercom technology exhibits a number of advantages over conventional RF technology, i.e. high level of security, ultra-wide usable bandwidth, frequency reuse (the same wavelength can be used for multiple links), mass, power consumption, and size reduction, immunity to radiation and electromagnetic interference, and absence of any frequency regulation [4]. In addition, the components, e.g. lasers, modulators, amplifiers, and photodetectors, for the implementation of free-space optical (FSO) links between satellites, and between a satellite and a ground station, have also reached a good level of maturity/reliability in harsh environments.

The advantage of lasercom technology in terms of size, weight, and power consumption is mainly due the lower level of diffraction suffered by an optical beam with respect to an RF beam. The wavelength λ utilized in the lasercom system is about 10^{-6} m, i.e. 10^4 times smaller than the typical operating wavelength of RF systems. Since the far-field beam divergence is equal to λ/D_{TX}, D_{TX} being the transmitter aperture diameter, the free-space laser communication system requires a transmitter aperture diameter 10,000 times smaller than RF communications to guarantee the same beam divergence [5].

However, from a technical point of view, the pointing and tracking system and the techniques for mitigating the detrimental effects due to cloud coverage and/or fog, are critical aspects in lasercom system design.

In the 1970s, NASA, Japan Aerospace Exploration Agency (JAXA), ESA, and several companies in the US, Europe, and Japan started their R&D activities on lasercom technologies [6–11]. After about 40 years of theoretical investigations and experiments carried out by several research groups all over the world, some optical links allowing data transfer at a bit rate of the order of tens/hundreds of Mb/s between satellites or space vehicles and a ground station have been successfully demonstrated in the last decade. Laser communication terminals for space-to-ground and inter-satellite communications have already reached the commercialization stage [12].

In this chapter, the field of optical links for intra- and inter-spacecraft communications is reviewed with a special emphasis on recent advances and future trends.

3.1 Fiber data links

Current trends in the space industry suggest that future satellites for telecommunications and Earth observation will produce or handle data up to several Tb/s.

Optical links seem to be the best technological solution to cover all these emerging needs [13]. They consist of a transmitter, a transmission channel (optical fiber), and a receiver. The transmitter converts the electrical digital data into a modulated optical beam, which propagates through the optical fiber and is converted back into the electrical domain at the receiver.

Telecom payloads will widely use the multi-beam technology, which offers much higher broadband capacity than typical wide-beam systems [14]. This approach implies the division of the satellite coverage area

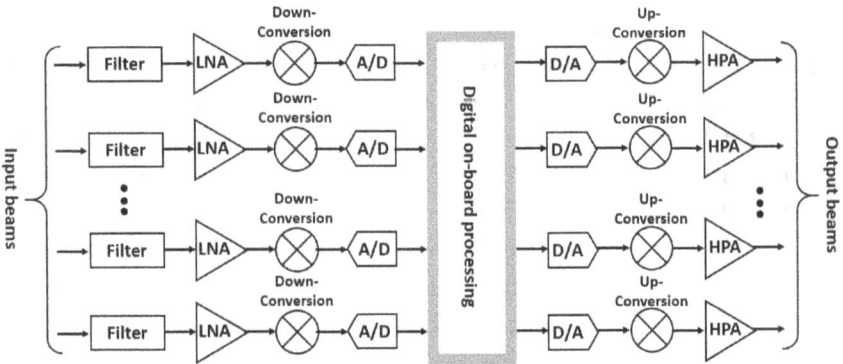

Fig. 3.1. Typical configuration of a digital telecom payload. LNA: low noise amplifier. HPA: high power amplifier.

into hundreds of cells, each served by a narrow spot beam having a bandwidth typically ≤1 GHz. Each beam is filtered, amplified, down-converted, digitalized, and then processed on-board in the digital domain, within the digital telecom payload (see Fig. 3.1). After processing, each beam is D/A converted, up-converted, and amplified before retransmission [15]. In the last few years, telecom missions have been migrating to the Ka-band, i.e. input beams at 30 GHz and output beams at 20 GHz.

This scenario allows us to envisage data traffic of about 5–10 Tb/s from the A/D converters to the digital signal processor (DSP) and from the DSP to the D/A converter.

Earth observation missions based on new active/passive instruments, demand innovative sub-systems for on-board data pre-processing and compression to enable, for example, the real-time availability of Earth images [16]. Such functionalities imply that the imager output is A/D converted at high-resolution and thus-generated data stream is transmitted to the on-board pre-processor at high speed. For example, a typical synthetic aperture radar antenna includes tens of sub-apertures, each connected to an A/D converter. The digital form of the signal generated by each sub-aperture has a bit rate of a few Gb/s, thus the data stream from the A/D converters to the on-board pre-processor has a bit rate up to 0.1 Tb/s.

To implement a bidirectional point-to-point channel, a transceiver including a transmitter and a receiver is placed at the two ends of the link.

Companies and research centers have developed opto-electronic trans-ceivers for high-speed connecting of several sub-systems on board space-crafts [17–20]. They are compliant with the very stringent requirements of the space missions, e.g. temperatures range from −40°C to +85°C, resist-ance to radiation up to a total dose of 1,000 Gy, resistance to vibration up to 25 g_{rms}. Most advanced transceivers for space applications include up to four transmitters and four receivers, which allow up to four bi-directional point-to-point communication channels. The bit rate for each channel can be up to 12.5 Gb/s and the point-to-point link length is of the order of 100 m [20]. Typically, a multi-fiber cable connects the transceivers at the ends of the link, as shown in Fig. 3.2. Thus, two optical fibers are used for each bi-directional channel.

In the first optical link for space applications, such as that mounted on board the SAMPEX satellite, directly modulated light emitting diodes (LEDs) operating at 850 nm were used for electrical/optical (E/O) trans-duction [21]. Although LEDs are low cost, low power, and very reliable devices, their use in the context of optical links limits the maximum bit

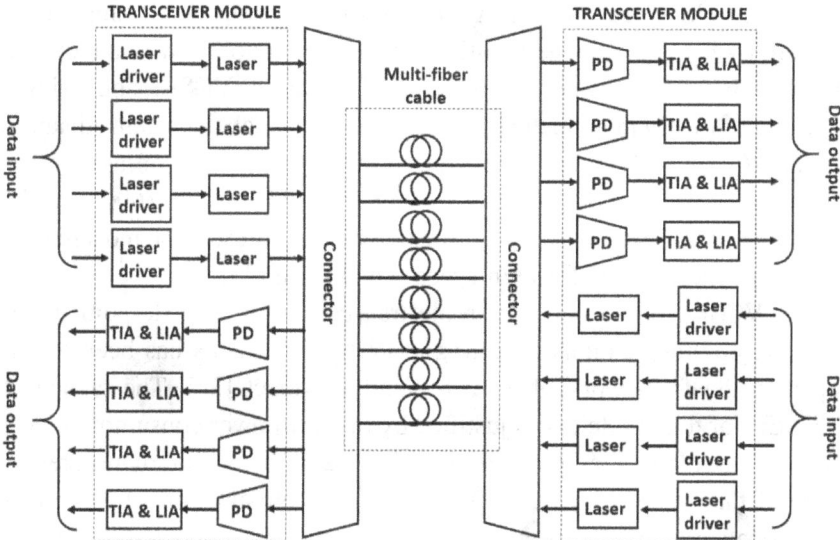

Fig. 3.2. Typical configuration of a 4 Tx + 4 Rx, transceiver for parallel optical data communications. PD: photodiode. TIA: transimpedance amplifier. LIA: limiting amplifier.

rate to about 100 Mb/s, which is not compliant with the current needs at system level. Thus, for more than two decades, directly modulated coherent semiconductor lasers, typically operating at cither 850 nm or 1330 nm, have been preferred to implement the optical transmitter.

The ESA used optical links for the first time in the Soil Moisture and Ocean Salinity (SMOS) Earth observation satellite [22]. For that mission, a directly modulated Fabry–Perot laser diode operating at 1,300 nm, which passed severe qualification test, including gamma and proton irradiation, thermal vacuum, and vibration, was selected for the E/O transduction [23].

Currently, the basic requirements for a semiconductor laser to be included in an opto-electronic transceiver for space applications are low cost, small size, low power consumption (E/O conversion efficiency $\geq 40\%$), high speed (direct modulation at a bit rate ≥ 10 Gb/s), and ease of coupling with the optical fiber. On the basis of these requirements, the interest in vertical-cavity surface emitting lasers (VCSELs) is quickly growing. VCSELs exhibit several features which are very useful in short-distance interconnections, such as high direct modulation bandwidth (supported bit rate up to 50 Gb/s in prototypes for terrestrial applications [24]), low threshold current (<1 mA), low power consumption, temperature sensitivity lower than edge-emitting lasers, easy coupling with the optical fiber, high density of integration (1D or 2D arrays of VCSELs on a single chip are routinely manufactured at low cost). Although the recent advances in the VCSEL technology have allowed the demonstration of laser covering a wide wavelength range extending from 650 nm to 1600 nm, the AlGaAs/GaAs lasers emitting at around 850 nm represent the most mature technology, especially for high-speed modulation [25].

A VCSEL-based opto-electronic transceiver for point-to-point intra-satellite communications operating at up to 3.125 Gb/s has been developed in the framework of an ESA-funded project [26]. The choice of VCSEL for that module was justified by the low power consumption/heat dissipation, the low temperature sensitivity, and the good radiation resistance of that laser. Commercial VCSEL-based opto-electronic transceivers for space applications operating at 850 nm are now available [27, 28].

For optical/electrical (O/E) transduction, p–i–n photodiodes are typically used in short-distance fiber-optic links because their bandwidth is compliant with the current application needs, and they have a responsivity higher than metal-semiconductor-metal photodiodes, and a cost lower than avalanche photodiodes. To make optical coupling with optical fiber easy, *p–i–n* photodiodes with a large photosensitive surface area are usually selected, but large sensitive area and high speed are conflicting requirements. Thus, a trade-off between the two key features of the device is strictly required. In this application context, III–V semiconductors (such as GaAs or InGaAs) *p–i–n* photodiodes are typically preferred to the Si *p–i–n* photodiodes because they are less sensitive to single event upsets, thanks to their thinner depletion region [29].

Each laser diode included in an opto-electronic transceiver needs an electronic driver circuit, which is an interface between the digital voltage signal, a differential current mode logic (CML) signal, and the signal driving the laser, which directly modulates the laser beam according to the non-return-to-zero on–off keying coding. The small photocurrent generated by the *p–i–n* photodiode is converted into a voltage by a transimpedance preamplifier (TIA). A limiting amplifier (LIA) is utilized to convert the preamplifier output into a CML digital output signal.

The laser drivers, TIAs, LIAs, and all other electronic devices/circuits are integrated in a rad-hard application-specific integrated circuit (ASIC) device. The opto-electronic components, i.e. the laser diodes and the *p–i–n* photodiodes, are hybridly integrated with the ASIC using the flip-chip technique.

The typical power consumption of an opto-electronic transceiver for space applications allowing four bidirectional point-to-point communication channels at 1–10 Gb/s is about 1 W. The transceiver volume is <1 cm^3 and its weight is only a few grams [30].

The transceivers placed at the two ends of a point-to-point fiber data link are usually connected by graded-index multimode optical fibers having a core diameter in the range 50–100 μm. This kind of fiber, having a modal bandwidth up to 1 GHz-km or more, is preferred to single-mode fibers, due to easier fiber/laser and fiber/photodiode optical alignment. Optical fibers are sensitive to gamma radiation especially because their attenuation notably increases when they are exposed to radiation. Radiation induced attenuation depends on two factors, i.e. the total irradiation dose and the radiation dose rate [31]. Assuming a total dose of 1000 Gy, a wavelength equal to 850 nm, and room temperature, the radiation induced attenuation of high-quality rad-hard fibers is <0.5 dB/km when the radiation dose rate is 0.06 Gy/h (typical background radiation level in space). If the dose rate increases up to 20 Gy/h (the dose rate corresponding to solar flare events) and the total dose is kept constant at 1000 Gy, the radiation induced attenuation of the same fibers is <15 dB/km [32]. Those values increase remarkably when the temperature drops down to about −20°C and decreases by more than one order of magnitude when the operating wavelength becomes 1300 nm [32]. For a point-to-point optical link operating at 850 nm, with a length of 100 m, we expect a radiation induced attenuation of the fiber cable <10 dB under all operating conditions.

Due to the continuous progress in silicon photonics, low-cost and low-power optical interconnections for board-to-board, chip-to-chip, and on-chip data transfer at a bit rate of the order of 10 Gb/s have been the topic of intense research effort for more than a decade. Although this technology is intended for terrestrial applications, its application in the space environment has already been studied especially for future on-board processors [33].

3.2 Fiber networks for distribution of high-purity analogue signals

The distribution of local oscillator signals in broadband telecom payloads, including complex multi-beam antennas, is becoming critical in terms of mass, complexity, and electromagnetic compatibility. Some Earth observation satellites need subsystems for distributing reference signals, which are immune to electromagnetic interference and preserve the signal purity.

One of the most attractive solutions to these needs is the use of optical fibers for distributing high-purity analogue signals. For example, in the SMOS satellite, due to their mechanical flexibility and immunity to electromagnetic interference, optical fibers were selected as the reference clock distribution to the RF receivers of a satellite radiometer that monitors the ocean salinity and the soil moisture over the land surfaces though L-band 2D interferometric radiometry [34].

In digital telecom payloads (see Fig. 3.1), frequency down- and up-conversions can be realized in two steps, e.g. from the Ka band to the intermediate frequency (IF) and from the IF to the baseband, and *vice versa*. The Ka band to IF or IF to Ka band conversion units uses a local oscillator signal, which is synthesized in the same units starting from an Ultra-Stable Reference Oscillator (USRO) signal at 10 MHz. The conversion IF–baseband (or baseband–IF) takes place in mixers that directly utilize a Master Local Oscillator (MLO) signal at about 800 MHz. In payloads based on this configuration, fiber networks have been proposed for the distribution to tens or hundreds of equipments of both the USRO at 10 MHz and the MLO at about 800 MHz, both generated in the electronic domain. The optical distribution network should not add phase noise to the RF signal. On the basis of system considerations, two different requirements for the USRO and the MLO distribution were identified, i.e. a phase noise floor at the output nodes of the distribution network around a −165 dBc/Hz for the USRO and −130 dBc/Hz for the MLO [35].

The basic configuration of the optical distribution network is shown in Fig. 3.3. The oscillator signal, i.e. USRO or MLO, is converted into the optical domain. The optical beam is amplified and split into tens or hundreds of signals, which are transmitted on fibers (length of the order of 100 m). At the output of each fiber, the O/E transduction at the input of each equipment allows us to reconstruct the oscillator signal. The use of an optical amplifier, typically an erbium doped fiber amplifier (EDFA), allows the optical power at the splitter input to be increased and thus the number of equipments.

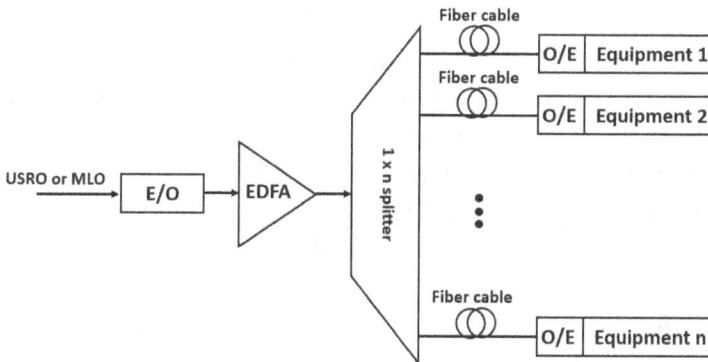

Fig. 3.3.   Typical configuration of a fiber network for distribution of high-purity analogue signals.

The phase noise power at the output of the O/E transduction units depends on the optical power at the input of that unit and the relative intensity noise (RIN) of the laser included in the E/O transduction unit both at the oscillator frequency and at low frequency [36]. Therefore, a low-RIN laser diode is the key element in the E/O transduction unit. Since the frequencies of both USRO and MLO are less than the bandwidth of the semiconductor laser for analogue applications, the E/O transduction is performed by the direct modulation of a laser diode. The stringent requirement on phase noise imposes the selection of a laser source with a very low RIN, which is typically a distributed feedback (DFB) laser having a RIN <−150 dB/Hz. The performance in terms of RIN of the VCSEL is currently non-compliant with the specific application [37].

The most common implementation of an O/E transduction unit is based on a photodiode and a low-noise transimpedance amplifier. Although all experimental results available in the literature suggest that this approach is appropriate for MLO distribution, it is non-compliant with the requirements on phase noise for USRO distribution. In the latter case, the only way to satisfy the requirements is the implementation of O/E transduction through an injection locked photo-oscillator, which includes a photodiode with an oscillator at the output. The oscillator locks itself onto the incoming RF signal and filters the signals far from the carrier, thus allowing a high level of the purity of the oscillator signal at the output of the O/E transduction unit.

By using a DFB laser with an optical power of 40 mW and a RIN of −160 dB/Hz directly modulated by an RF signal having a power of 10 dBm, and an optical amplifier providing an output power up to 18 dBm, the distribution of the MLO to more than 256 equipments has been demonstrated with the required level of phase noise. The same components allow the distribution of a USRO to 64 equipments with appropriate purity only if injection locked photo-oscillators are used for the O/E transduction.

The distribution of local oscillator signals at a frequency greater than a few GHz implies the implementation of the E/O transduction unit through a low-RIN DFB laser diode and an electro-optic modulator. The distribution of two local oscillator signals at 12 GHz and 16 GHz by a fiber network has been demonstrated in Ref. [38] by realizing E/O transduction using laser external modulation and O/E transduction through a photodetector. The achieved phase noise floor at the network output nodes is −145 dBc/Hz.

As will be discussed in Chapter 4, the frequency down- and up-conversion can be performed directly in the optical domain. In this case, the local oscillator optical signal is obviously distributed through a fiber network [38, 39].

In the late 1970s, the NASA Jet Propulsion Laboratory developed the first network for reference signal distribution with ultra-high phase stability within the complex of deep space stations located in Goldstone, CA, USA [40, 41]. The complex is part of the NASA Deep Space Network, which is responsible for commanding, tracking, and monitoring spacecraft exploring deep space, and has a number of stations, each including a large dish antenna, a high power microwave transmitter, and an ultra-low noise receiver. The developed network allowed the distribution of the reference signal generated by a high precision hydrogen maser to all stations in the Goldstone complex [42].

About two decades later, an optical fiber was used on board the NASA Space Shuttle to transmit a reference signal at 5.3 GHz [43]. The fiber selection was based on its superiority with respect to coaxial cable in terms of weight, resistance to radiation, and thermal coefficient of delay (TCD), quantifying the effect of temperature drift on the signal propagation delay (the TCD of fibers is less than a few ppm/°C while the TCD of the coaxial cable is >40 ppm/°C). The Shuttle Radar Topography Mission (SRTM) aiming at acquiring a digital topographic map of 80% of the Earth landmass by employing both a C-band and an X-band synthetic aperture radar (SAR) was launched [43]. Since the topographic map was obtained by using an interferometric technique, two sets of SAR antennas were utilized. These sets, each including one X-band antenna and one C-band antenna, were located at a distance of 60 m using a mast fixed on the Space Shuttle (see Fig. 3.4).

The technique exploited for topographic map creation required a stable reference signal at 5.3 GHz transmitted from the Shuttle to the end of the mast where two receiving antennas were located. The transmission was performed by a 1.55 μm single-mode point-to-point microwave fiber-optic

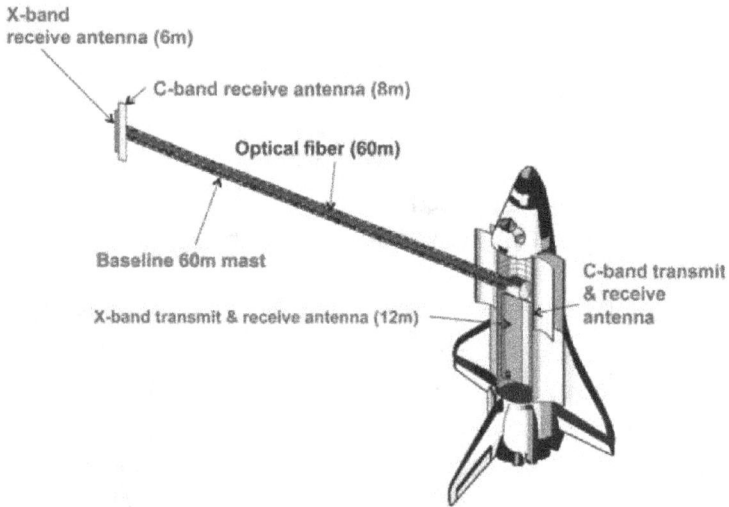

Fig. 3.4.   SAR antennas on the Shuttle Endeavour during the SRTM mission.
Source: Reprinted from Ref. [43], with the permission of IEEE.

link (see Fig. 3.5), having a crucial role for the mission's success. The transmitter module included a single-mode DFB CW laser diode (RIN approximately −150 dB/Hz) generating the optical carrier which was modulated by a LiNbO$_3$ Mach–Zehnder modulator with a bandwidth of 18 GHz. Before transmission, the optical beam passed through an optical phase shifter that compensated the eventual phase error between the reference signal and the signal received by the antennas at the end of the mast. The O/E transduction was based on a *p–i–n* photodiode with a responsivity of 0.8 A/W, a bandwidth of 12 GHz, and a spectral range from 1100 nm to 1650 nm [44]. Due to the very stringent requirements on the phase stability of the reference signal, a single mode fiber with a very low TCD (< 1 ppm/°C) was used [45].

The transmitters and the receivers were protected from radiation by an aluminum shielding having a thickness of approximately 2.5 mm. The fiber cable was also shielded from radiation.

As shown in Fig. 3.5, the optical phase shifter was driven by a phase comparator, which compares the phase of the reference signal and the phase of the fiber-propagating optical beam. A partially reflecting mirror

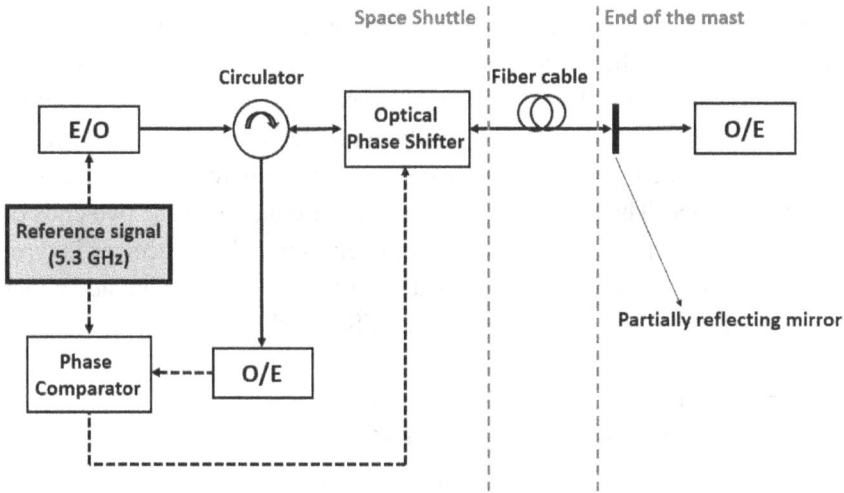

Fig. 3.5.   Block diagram of the microwave fiber-optic link used in the SRTM.

was placed at the end of the fiber cable to partially back-reflect the optical beam from the end of the mast towards the Space Shuttle. The back-reflected beam was O/E transduced within the Space Shuttle by a *p–i–n* photodiode and the resulting microwave signal sent to the phase comparator. If the phase comparator detects a phase error due to, for example, drifts in the temperature outside the mast, the transmitted optical signal is appropriately phase shifted by the optical phase shifter [46]. This active stabilization system assures that the two sets of antennas receive the reference signal with a maximum phase change <3°.

The three-port fiber circulator placed at the transmitter output allows the optical beams traveling in opposite directions to be separated and exhibits an insertion loss <1.3 dB, a minimum peak isolation of 45 dB, and an optical return loss ≤−50 dB.

3.3 Wireless data links

Data transfer at a bit rate <1 Mb/s between units on board a spacecraft, which are located at a distance less than a few meters, can be enabled by optical wireless links. This approach, which allows significant mass, size, and power saving, has been studied for more than two decades at the National Institute for Aerospace Technique of Spain and at the Applied Physics Laboratory at Johns Hopkins University [47, 48]. A unidirectional communication channel can be easily implemented through an emitter that performs the E/O transduction and a receiver where the O/E transduction takes place. The emitter and the receiver are placed at the two ends of the link, but the line-of-sight between the emitter and the receiver is typically unsecured. The emitter, having a size of a few cm^3, can include either an infrared emitting diode (IRED) or a VCSEL, while the receiver is based on a *p–i–n* silicon photodiode.

This technology was evaluated inside a Russian spherical capsule with a diameter of about 2 m, named FOTON-M3 (see Fig. 3.6), which carried a payload of around 40 European experiments and was launched

Fig. 3.6.   FOTON-M3 Russian spherical capsule.

Source: Reprinted from Ref. [47], with the permission of IEEE.

on September 14, 2007 [49]. Two optical links at 950 nm operating at a bit rate of 62.5 kb/s were installed on board the capsule, which was in low Earth orbit (LEO) for 12 days. The measured bit error rate was in the range from 10^{-7} to 10^{-4}, depending on the relative position of the transmitter and the receiver [50].

Optical wireless data links intra-satellite communications seem useful only in the context of micro/nanosatellites [51], and currently their bit rate and bit error rate are very far from those assured by fiber links.

3.4 Spacecraft-to-ground and satellite-to-satellite links

Lasercom links can be used to transmit data form LEO, geostationary orbit (GEO), and deep-space spacecraft to ground and vice versa.

The first demonstrations of laser communications between a satellite in GEO/LEO and a ground station were achieved starting from the mid-1990s [52]. Obviously, the development of high-speed data links from deep-space is significantly more complex. In 2003, NASA initiated the Mars Laser Communication Demonstration (MLCD) project aiming at demonstrating the first lasercom system transmitting data from Mars. The flight terminal of the optical link would have flown aboard the Mars Telecom Orbiter (MTO) [53]. On July 21, 2005, NASA announced the cancellation of MTO mission and thus the MLCD project was terminated [54]. Very recently, in the framework of the Lunar Laser Communication Demonstration (LLCD) mission, NASA demonstrated an FSO link from a satellite, the Lunar Atmosphere and Dust Environment Explorer (LADEE), orbiting around the Moon's equator, and three ground stations [55]. The distance between the link terminals is about 400,000 km and the maximum achieved data rate is 622 Mb/s.

The basic block diagram of a spacecraft-to-ground bidirectional lasercom link is shown in Fig. 3.7. It includes a flight transceiver, an Earth transceiver, and an optical channel. Lasercom systems need a clear line of sight between the two ends of the link. Thus, the ground station transmits a laser beacon towards the flight terminal to guide the downlink beam. The beacon beam is detected by a dedicated photodetector. A data stream with a low bit rate can also be transmitted from the ground station to the spacecraft by modulating the laser beacon.

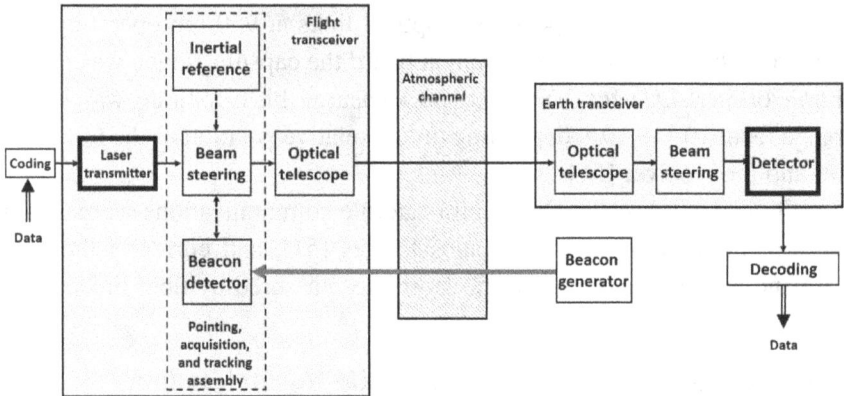

Fig. 3.7.   Spacecraft-to-ground bidirectional FSO link.

The function of pointing, acquisition, and tracking assembly is the detection of the incoming beacon signal and the very accurate pointing of the laser beam towards the receiver within the ground station.

The laser transmitter output is a beam confined in an optical fiber. The telescope in the flight transceiver expands the beam with a diameter of about 10 μm to a much larger diameter, which is usually of the order of centimeters. The telescope in the Earth transceiver focuses the received beam onto the core of the fiber that couples the beam to the photodetector.

Optical beams passing through the free-space optical channel experience several atmospheric effects, which can degrade the transmitted signals [56, 57]. The most critical atmospheric effects are atmospheric attenuation, atmospheric scattering mainly due to snow, clouds, fog, or rain, atmospheric turbulence, and, finally, atmospheric radiance, which affects the signal-to-noise ratio of the received beam. Further details on all these contributions will be provided in Section 4.4.

Lasercom FSO links have been proved to be very attractive for data transfer between two satellites, which are both in a GEO/LEO orbit, or one in a GEO orbit and the other in an LEO orbit. An important advantage of satellite-to-satellite links is the lack of atmosphere between the two transceivers, while a design challenge for those systems is that neither of the lasercom host platforms is motionless. The ESA semiconductor laser intersatellite link experiment (SILEX) demonstrated a bidirectional link between an Earth observation satellite (SPOT-4) and a GEO satellite (ARTEMIS) [58, 59]. The bit rate of the data stream from SPOT-4 to ARTEMIS is 50 Mb/s. ARTEMIS transmitted to Earth (Tolouse ground station) the images optically received by SPOT-4 *via* Ka-band RF signals. The distance between the two satellites is of the order of 40,000 km. To establish a clear line of sight between the link terminals, the laser terminal on ARTEMIS emits a beacon beam generated by a combination of several high-power (500 mW) laser diodes. Since the uncertainty on the ARTEMIS attitude is 1,700 μrad, that satellite scans, with a beacon having a beam divergence of 700 μm, the 1,700 μrad uncertainty cone. When SPOT-4 detects the beacon, it sends a communication beam to ARTEMIS that stops the beacon scan. Then the two link terminals, on ARTEMIS and SPOT-4 track each other, optimizing their alignment. After this step, the laser terminal on ARTEMIS switches off the beacon and switches on its communication beam. This acquisition sequence is summarized in Fig. 3.8.

The SILEX link statistics counts more than 1,800 sessions, with a failure rate of 4% and an accumulated link duration of about 380 h [59].

Several years later, a lasercom FSO link, operating up to 5.6 Gb/s, between two satellites (TerraSAR-X and Near Field Infrared Experiment,

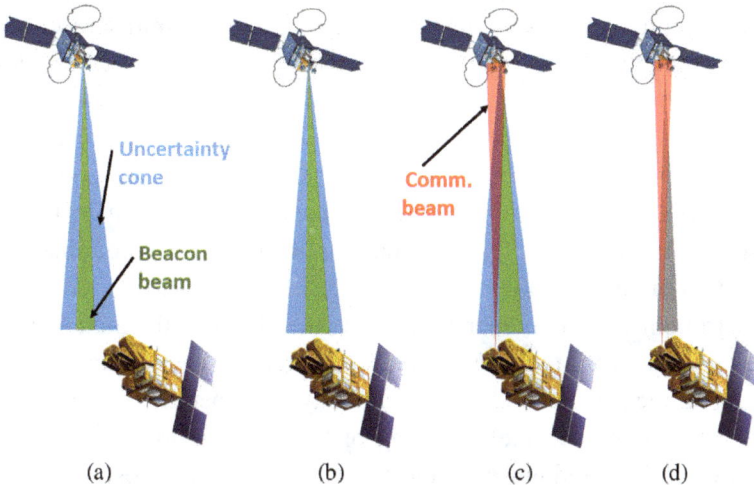

Fig. 3.8. Acquisition sequence. (a) The ARTEMIS beacon scans the uncertainty cone. (b) SPOT-4 detects the beacon. (c) SPOT-4 sends a communication beam to ARTEMIS that stops the beacon scan. (d) ARTEMIS switches off the beacon and switches on its communication beam.

NFIRE) in LEO orbit was demonstrated [60]. In this case the distance between the two link terminals is about 5,000 km and the accumulated link duration is about 4.5 h. A space-to-ground bidirectional optical communication link at 5.6 Gb/s has also been demonstrated between the NFIRE satellite and an ESA ground station in Tenerife (Spain) [61].

3.5 Atmospheric channel

The performance of satellite/ground lasercom links are strongly influenced by the features of the atmospheric channel. In particular, the most critical effect, usually limiting the link performance, is due to atmospheric turbulence. In fact, that random fluctuations of the atmosphere refractive index induces a degradation of the quality of the laser beam carrying the data stream, with a consequent increase in the link bit error rate. Absorption and scattering are the two loss sources for the transmitted beam while the received signal-to-noise ratio is degraded by the background radiation from the Sun, the Moon, or a planet.

Air molecules and atmospheric aerosols, i.e. particles with a diameter ranging from 0.002 μm to 100 μm that are suspended in the atmosphere, cause both attenuation and scattering. Loss due to both these physical effects strongly depends on the wavelength of the transmitted laser beam. Starting from the 1970s, a number of software packages have been developed to evaluate and predict the absorption/scattering effects in several weather conditions and over a wide wavelength range [62]. For example, the transmittance spectrum of the atmosphere for a space-to-ground link, as calculated by one of these software packages is shown in Fig. 3.9. These results show that the atmospheric transmittance is about 0.8–0.9 at both 1,064 nm and 1,550 nm.

Astronomical bodies such as a planet, the Moon, or the Sun, which are in line of sight with the transmitter, are sources of background noise for the ground receiver. In addition, during daytime operations, the sky radiance, which is due to the scattering of the Sun's irradiance, is another source of background noise for the receiver. One of the most critical conditions is when the Sun is within the receiver field-of-view.

The atmosphere refractive index, which depends on air temperature/density, randomly varies in space and time because of the heating of the Earth surface caused by sunlight. The random nature of the atmosphere

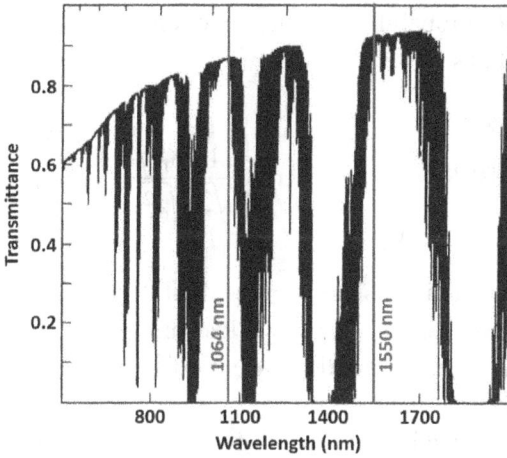

Fig. 3.9. Transmittance spectrum of the atmosphere for a space-to-ground link at sea level. Zenith angle = 0°, mid-latitude, and visual range = 23 km.

Source: Adapted from Ref. [63].

refractive index seriously affects the phase, intensity, and direction of the transmitted optical beam.

Signal intensity fluctuations at the receiver, usually called signal scintillation, have a direct impact on the bit error rate of the link. It is quantified by the scintillation index, which is given by [64]

$$\sigma_I^2 = \frac{\langle I^2 \rangle}{\langle I \rangle^2} - 1, \tag{3.1}$$

where I is the received optical intensity, and the symbol $\langle \, \rangle$ indicates the average. By using the Rytov method, the intensity probability density function can be written as [64]

$$p_I(I) = \frac{\sigma_I}{\sqrt{2\pi}} \exp\left[\frac{\left(\ln\left(\frac{I}{\langle I \rangle}\right) + \frac{1}{2}\sigma_I^2 \right)^2}{2\sigma_I^2} \right]. \tag{3.2}$$

The intensity probability density function allows to be expressed the link bit error rate as

$$\text{BER} = \int\limits_{0}^{+\infty} \frac{1}{2} p_1(I) \operatorname{erfc}\left(\frac{1}{\sqrt{2}} \frac{R_{pd} L_T A_{pd} I}{\sqrt{\sigma_1^2} + \sqrt{\sigma_0^2}}\right), \qquad (3.3)$$

where I is the optical intensity when the digit "1" is transmitted, R_{pd} is the photodetector responsivity, L_T is total loss of the link, A_{pd} is the photodetector area, and σ_1^2 and σ_0^2 are the current noise powers when the digit '1' or '0' is transmitted.

An accurate selection of the site for the ground station can mitigate the optical link degradation due to the atmospheric channel. Moreover, signal scintillation can be overcome by spatial transmitter diversity [65], i.e. the transmission of the same data stream by several parallel transmitters with an appropriate space separation with a consequent beam propagation through statistically independent atmospheric channels. Finally, the detrimental effects of the atmospheric turbulence can be reduced by using appropriate error-correcting codes [66], innovative modulation formats, e.g. Multimode-Differential Phase Shift Encoding [67], or adaptive optics [68], which consists of measuring and correcting the wave-front deformations of the received beam by applying the inverse deformation through, for example, deformable mirrors.

3.6 Laser transmitters and receivers

The key components of the transmitters utilized in lasercom systems are the semiconductor lasers, the diode-pumped Nd:YAG solid state lasers at 1064 nm, the electro-optic modulators, and the optical amplifiers with very high output power. Depending on the requirements at system level imposed by the space mission target, the most appropriate configuration of the transmitter is selected and consequently, the components are chosen.

In the NASA mission LLCD, a master oscillator power amplifier (MOPA) transmitter has been included in the flight terminal [69]. The block diagram of the transmitter is shown in Fig. 3.10. The distributed

Data

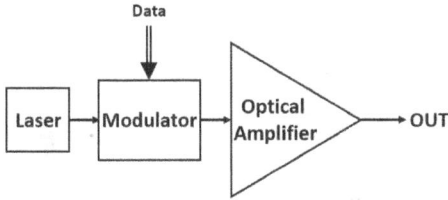

Fig. 3.10.   Block diagram of the MOPA transmitter.

feedback laser generates a continuous-wave beam at 1,550 nm, which is modulated by a Mach–Zehnder electro-optic modulator. The modulated beam is then amplified by a two stage Erbium-doped fiber amplifier (EDFA). At the amplifier output, the average optical power is 500 mW. The maximum bit rate supported by the MOPA transmitter is 622 Mb/s and the modulation format is pulse position modulation (PPM).

The laser transmitter included in the two transceivers utilized in the ESA SILEX mission are both based on single-mode GaAlAs semiconductor lasers. The laser diode on the SPOT-4 satellite operates at 847 nm, emits an optical power of 60 mW, and is modulated at 50 Mb/s, while the laser on the ARTEMIS satellite emits an optical power of 37 mW at 819 nm and supports a bit rate of 2 Mb/s [69]. The modulation format is on–off keying (OOK).

The optical links between TerraSAR-X and NFIRE and between NFIRE and the ESA ground station in Tenerife use a more complex modulation format than OOK or PPM, i.e. binary phase-shift keying (BPSK), which also allows the link operation to be kept when the Sun is within the receiver field-of-view [70]. To implement the BPSK transmitter, a laser source generating a beam with very good spectral and spatial features, e.g. linewidth of the order of 0.1 MHz, is required. Semiconductor lasers do not usually comply with these requirements. Therefore, the BPSK transmitters included in the transceivers on board TerraSAR-X and NFIRE are implemented through a diode-pumped Nd:YAG laser emitting at 1,064 nm and an electro-optic phase modulator.

The simplest configuration of the receiver, based on direct detection [71], is shown in Fig. 3.11. It includes an optical filter, a photodiode, an electronic amplifier, and a sampling/decision circuit. The filter is used to reduce background radiation at the photodiode input and it is centered at

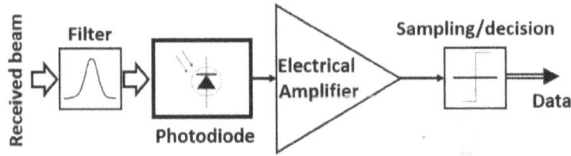

Fig. 3.11.   Receiver configuration based on the direct detection.

the link operating wavelength, e.g. 1,550 nm. The photodiode is the key element of the receivers and, typically, either an avalanche photodiode (APD) or photon-counting detector is used for this application. Since photon-counting detectors are usually bulky and power consuming, especially because they operate at cryogenic temperatures, they are only utilized in the ground stations.

The receiver performance, in terms of bit error rate, can be improved if an optical pre-amplifier, e.g. an EDFA, is placed before the filter. The basic requirements for the amplifier are low noise figure and high gain.

Theoretical calculations suggest that the best receiver sensitivity can be obtained using a coherent receiver [71], whose block diagram is shown in Fig. 3.12. In this configuration, the received optical beam is mixed with an optical beam generated by a local laser oscillator within the receiver. The spectrum of the electrical signal generated by the photodiode is centered at the frequency resulting from the difference between the operating frequency of the local laser oscillator and the central frequency of the received optical beam. This receiver configuration allows quite easy detection of phase modulated optical beams.

In the LLCD mission, direct detection is used at the ground station receiver. The photodetector is a 12 pixel array of WSi superconducting nanowire single photon detectors operating at cryogenic temperature [72, 73]. The array consists of multiple superconducting nanowires acting as single photon detectors. The nanowires are independently biased.

The simplest detector configuration, i.e. direct detection without any optical pre-amplification, was utilized in the SILEX mission [74]. The selected detector is a silicon APD. In the TerraSAR-X–NFIRE optical link, a more sophisticated detection approach, based on a coherent configuration, was chosen [75]. The laser communication terminal used on board both TerraSAR-X and NFIRE is shown in Fig. 3.13. It transmits an

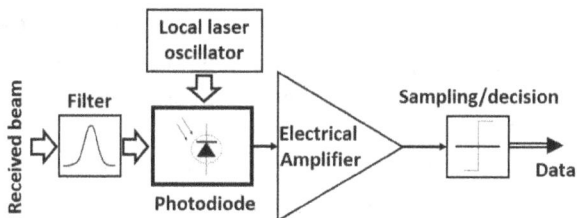

Fig. 3.12.   Receiver configuration based on the coherent detection.

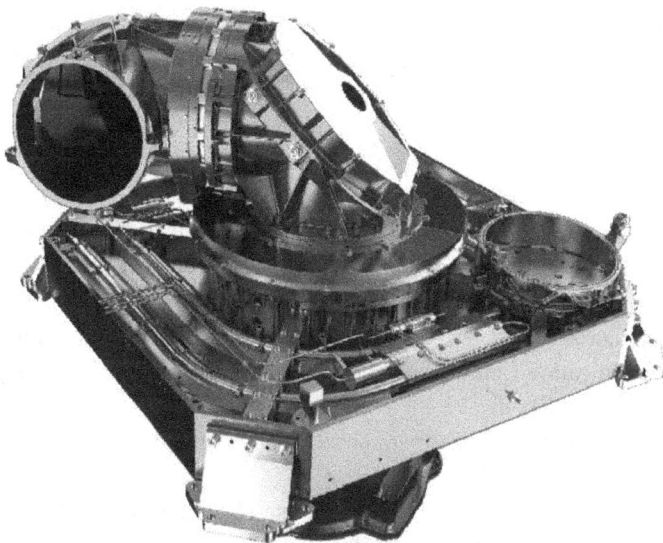

Fig. 3.13.   Laser communication terminal on board of TerraSAR-X and NFIRE.
Source: Reprinted from Ref. [12], with the permission of IEEE.

optical power of 0.7 W at 1064 nm (Nd:YAG laser) and supports a bit rate of 5.6 Gb/s.

References

[1]   K. A LaBel, C. J. Marshall, P. W. Marshall, P. J. Luers, R. A. Reed, M. N. Ott, C. M. Seidleck and D. J. Andrucyk (1998). On the suitability of fiber optic data links in the space radiation environment: a historical and scaling technology perspective, *Proceedings of the 1998 IEEE Aerospace Conference*, vol. 4, pp. 421–434, Aspen, CO, USA, March 21–28.

[2] M. N. Ott, F. LaRocca, W. J. Thomes, R. Switzer, R. Chuska and S Macmurphy (2008). Applications of optical fiber assemblies in harsh environments: The journey past, present, and future, Proceedings of SPIE, vol. 7070, 707009.

[3] N. Karafolas, J. M. P Armengol and I. Mckenzie (2009). Introducing photonics in spacecraft engineering: ESA's strategic approach, 2009 IEEE Aerospace conference, Big Sky, MT, USA, March 7–14.

[4] H. Hemmati (2009). Introduction," in *Near-Earth Laser Communications*, H. Hemmati, (Ed.), CRC Press, Florida.

[5] H. Hemmatia and D. Caplan (2013). Optical Satellite Communications, in *Optical Fiber Telecommunications*, I. P. Kaminow, T. Li and A. E. Willner (Eds.), VIB, Elsevier.

[6] Hughes Aircraft Co. (1964). Deep space optical communications systems study, NASA, CR-73.

[7] J. H. McElroy, N. McAvoy, E. H. Johnson, J. J. Degnan, F. E. Goodwin, D. M. Henderson, T. A. Nussmeier, L. S. Stokes, B. J. Peyton and T. Flattau (1977). CO_2 laser communication systems for near-earth space applications, *Proceedings of the IEEE*, vol.65, pp. 221,251.

[8] W. Reiland, W. Englisch and M. Endemann (1986). Optical Intersatellite Communication Links: State of CO_2 Laser Technology, *Proceedings SPIE*, vol. 616, pp. 69–76.

[9] M. Ross (1988). The history of space laser communications, *Proceedings of SPIE*, vol. 885(2).

[10] R. Dumas and B. Laurent (1990). System test bed for demonstration of the optical space communications feasibility, *Proceedings of SPIE*, vol. 1218, p. 398.

[11] Y. Arimoto, M. Toyoshima, M. Toyoda, T. Takahashi, M. Shikatani and K. Araki (1995). Preliminary result on laser communication experiment using ETS-VI, *Proceedings of SPIE*, vol. 2381, pp. 151–158.

[12] S. Seel, H. Kämpfner, F. Heine, D. Dallmann, G. Mühlnikel, M. Gregory, M. Reinhardt, K. Saucke, J. Muckherjee, U. Sterr, B. Wandernoth, R. Meyer and R. Czichy (2011). Space to ground bidirectional optical communication link at 5.6 Gbps and EDRS connectivity outlook, IEEE Aerospace Conference, Big Sky, MT, USA March 5–12.

[13] C. Ciminelli, F. Dell'Olio, M. N. Armenise, F. Iacomacci, F. Pasquali and R. Formaro (2012). Design and optimization of a fiber optic data link for new generation on-board SAR processing architectures, in *International Conference on Space Optical Systems and Applications (ICSOS) 2012*, Ajaccio, Corsica, France, October 9–12.

[14] T. M. Braun (2012). *Satellite Communications Payload and System*. Wiley & Sons, New Jersey.

[15] M. Sotom, B. Bénazet, A. Le Kernec and M. Maignan (2009). Microwave photonic technologies for flexible satellite telecom payloads, European Conference on Optical Communication, Wien, Austria, September 20–24.

[16] R. Trautner and R. Vitulli (2010). Ongoing developments of future payload data processing platforms at ESA, in *Second International Workshop on On-Board Payload Data Compression (OBPDC)*, Toulouse, France, October 28–29.

[17] P. Karioja and M. Karppinen (2012). VTT SolidOpto — High-Speed Data Links, CERN, Geneva, Switzerland, January 17.

[18] M. Pez, F. Quentel, G. Barbary, C. Claudepierre and C. Hartmann (2005). High performance mil/aero fiber optic transceiver, *IEEE Conference on Avionics Fiber-Optics and Photonics*, Sheraton Bloomington, Minneapolis, MN, USA, September 20–22.

[19] Space Photonics, FireFibe HMP series transceiver modules, http://www. spacephotonics.com/.

[20] C. Tabbert (2012). Non hermetic fiber optic transceivers for space applications, *International Conference on Space Optics*, Ajaccio, France, October 9–12.

[21] P. W. Marshall, C. J. Dale, K. A. LaBel and M. C. Flanegan (1996). Space radiation effects in high performance fiber optic data links for satellite data management, *Proceedings of the IEEE Aerospace Applications Conference*, Aspen, CO, USA, February 3–10.

[22] K. Kudielka, F. J. Benito-Hernández, W. Rits and M. Martin-Neira (2010). Fibre optics in the SMOS mission, *International Conference on Space Optics*, Rhodes, Greece, October 4–8.

[23] Press release, Modulight to deliver world's first telecommunication laser diodes for space application, http://www.modulight.com/press-releases/press-news-22/.

[24] D. Kuchta, A. Rylyakov, C. Schow, J. Proesel, F. Doany, C. W. Baks, B. Hamel-Bissell, C. Kocot, L. Graham, R. Johnson, G. Landry, E. Shaw, A. MacInnes and J. Tatum (2013). A 56.1Gb/s NRZ Modulated 850nm VCSEL-based optical link, *Optical Fiber Communication Conference*, paper OW1B.5, Anaheim, CA, USA, March 17–21.

[25] A. Mutig (2011). *High Speed VCSELs for Optical Interconnects*. Springer, New York.

[26] V. Heikkinen, T. Alajoki, E. Juntunen, M. Karppinen, K. Kautio, J.-T. Makinen, J. Ollila, A. Tanskanen, J. Toivonen, R. Casey, S. Scott, W. Pintzka, S. Theriault and

I. Mckenzie (2007). Fiber-optic transceiver module for high-speed intrasatellite networks, *Journal of Lightwave Technology*, vol. 25, pp. 1213–1223.

[27] See the webpage http://www.radiall.com/products/active-optics

[28] C. Tabbert (2014). Chip scale package fiber optic transceiver integration for harsh environments, *International Conference on Space Optics*, Tenerife, Spain, October 7–10.

[29] C. Barnes, M. Ott, A. Johnston, K. LaBel, R. Reed, C. Marshall and T. Miyahira (2002). Recent photonics activities under the NASA electronic parts and packaging (NEPP) program, *Proceedings SPIE*, vol. 4823, pp. 189–204.

[30] Datasheets of the TS-HMP Series of optical transceivers by Space Photonics, the DLR-xx-250-yQ-Pz and the DLT-xx-250-yQ-Pz transceivers by D-lightsys, and the X80-Q Fury SMT Quad Transceiver by Ultra Communications.

[31] C. DeCusatis (2013). *Handbook of Fiber Optic Data Communication: A Practical Guide to Optical Networking*. Accademic Press, New York.

[32] S. Thériault (2006). Radiation effects on COTS laser-optimized graded-index multimode fibers exposed to intense gamma radiation fields, *Proceedings SPIE*, vol. 6343, 63431Q.

[33] E. Grivas, E. D. Kyriakis-Bitzaros, G. Halkias, S. G. Katsafouros, G. Morthier, P. Dumon, R. Baets, T. Farell, N. Ryan, I. McKenzie and E. Armadillo (2008). Wavelength division multiplexing based optical back-plane with arrayed waveguide grating passive router, *Journal of Optical Engineering*, vol. 47, 025401, 2008.

[34] R. Palacio, F. Deborgies and P. Piironen (2010). Optical distribution of microwave signals for Earth Observation satellites, *IEEE Topical Meeting on Microwave Photonics (MWP)*, Montreal, QC, Canada, October 5–9.

[35] B. Bénazet, M. Sotom, M. Maignan and J. Berthon (2004). Optical distribution of local oscillators in future telecommunication satellite payloads, *International Conference on Space Optics*, Toulouse, France, March 30–April 2.

[36] M.-B. Bibey, F. Deborgies, M. Krakowski and D. Mongardien (1999). Very low phase noise optical links: Experiments and theory, *IEEE Transactions On Microwave Theory and Techniques*, vol. 47, pp. 2257–2262.

[37] S. Azaizia, K. Saleh, O. Llopis and A. Rissons (2012). Evaluation of low cost solutions for the transmission through optical fiber of low phase noise OCXO signals, *IEEE International Frequency Control Symposium (FCS)*, Baltimore, MD, USA, May 21–24.

[38] D. Yap, W. W. Ng, H.T.M. Wang,; K. R. Sayyah and D. L. Persechini (2000). RF-photonic links for local-oscillator distribution and frequency

conversion, IEEE International Conference on Phased Array Systems and Technology, pp. 371–374, Dana Point, CA, USA, May 21–25.

[39] M. Sotom, B. Bénazet, A. Le Kernec and M. Maignan (2009). Microwave photonic technologies for flexible satellite telecom payloads, European Conference on Optical Communication, Vienna, Austria, September 20–24.

[40] L. E. Primas, R. T. Logan and G. F. Lutes (1989). Applications of ultra-stable fiber optic distribution systems, 43rd Annual Symposium on Frequency Control, Denver, CO, USA, May 31–June 2.

[41] L. E. Primas, R. T. Logan, G. F. Lutes and L. Maleki (1990). Distribution of ultra-stable frequency signals of fiber optic cable, IEEE MTT-S International Microwave Symposium, Dallas, TX, May 8–10.

[42] M. Calhoun, H. Shouhua and R. L. Tjoelker (2007). Stable photonic links for frequency and time transfer in the deep-space network and antenna arrays, *Proceedings of the IEEE*, vol. 95, pp. 1931–1946.

[43] K. Y. Lau, G. F. Lutes and R. L. Tjoelker (2014). Ultra-Stable RF-Over-Fiber Transport in NASA Antennas, Phased Arrays and Radars, *Journal of Lightwave Technology*, vol. 32, pp. 3240–3451.

[44] D. N. Horwitz (2001). 60-m delay-stabilized microwave fiber optic link for the STS-99 Shuttle Radar Topography Mission (SRTM), *Proceedings SPIE*, vol. 4216, 218.

[45] Y. Shuto, F. Yamamoto and Y. Takeuchi (1986). High-temperature stability of optical transmission properties attained by liquid-crystal polymer jacket, *Journal of Lightwave Technology*, vol. LT-4, pp. 614–618.

[46] D. A. McWatters, G. Lutes, E. R. Caro and M. Tu (2001). Optical calibration phase locked loop for the shuttle radar topography mission, *IEEE Transactions On Instrumentation and Measurement*, vol. 50, pp. 40–46..

[47] I. Arruego *et al.* (2009). OWLS: A ten-year history in optical wireless links for intra-satellite communications, *IEEE Journal on Selected Areas in Communications*, vol. 27, pp. 1599–1611.

[48] R. C. Meitzler, W. Schneider and R. Conde (2005). IRCOMM: Spacecraft free-space optical bus development, IEEE Aerospace Conference, pp. 1583–1588, Big Sky, MT, USA, March 5–12.

[49] See the web page http://www.russianspaceweb.com/foton.html

[50] H. Guerrero (2011). Optoelectronic COTS Technologies for Planetary exploration, Europlanet Meeting, Helsinki, May 26.

[51] M. R. Mughal, L. M. Reyneri and D. Del Corso (2012). Implementation of Inter and Intra Tile Optical Data Communication for Nano-Satellites, International Conference on Space Optics, Ajaccio, France, October 9–12.

[52] Y. Arimoto, M. Toyoshima, M. Toyoda, T. Takahashi, M. Shikatani and K. Araki (1995). Preliminary result on laser communication experiment using (ETS-VI), *Proceedings SPIE*, vol. 2381, 151.

[53] S. F. Franklin, J. P. Slonski, S. Kerridge, G. Noreena, S. Townes, E. Schwartzbaum, S. Synnott, M. Deutsch, C. Edwards, A. Devereaux, R. Austin, B. Edwards, J. J. Scozzafav, D. M. Boroson, W. T. Roberts, A. Biswas, A. D. Pillsbury, F. I. Khatri, J. Sharma and T. Komarek (2004). The 2009 Mars Telecom Orbiter mission, Proceedings of the IEEE Aerospace Conference, Big Sky, MT, USA, March 6–13.

[54] A. Biswas, D. Boroson and B. Edwards (2006). Mars laser communication demonstration: What it would have been, *Proceedings SPIE*, vol. 6105, 610502.

[55] D. M. Boroson, B. S. Robinson, D. V. Murphy, D. A. Burianek, F. Khatri, J. M. Kovalik, Z. Sodnik and D. M. Cornwell (2014). Overview and results of the Lunar Laser Communication Demonstration, Proceedings of SPIE, vol. 8971, 89710S.

[56] L. C. Andrews and R. L. Phillips (2005) *Laser Beam Propagation through Random Media*, 2nd Edn. SPIE Press, Washington.

[57] J. C. Ricklin, S. M. Hammel, F. D. Eaton and S. L. Lachinova (2010). Atmospheric channel effectson free-space laser communication, in *Free-Space Laser Communications*, A. K. Majumdar and J. C. Ricklin (Eds.), Springer, New York.

[58] T. T. Nielsen and G. Oppenhaeuser (2002). In orbit test result of an operational intersatellite link between ARTEMIS and SPOT4, SILEX, *Proceedings SPIE*, vol. 4635.

[59] Z. Sodnik, B. Furch and H. Lutz (2010). Optical Intersatellite Communication, *IEEE Journal of Selected Topics in Quantum Electronics*, vol. 16, pp. 1051–1057.

[60] R. Fields, C. Lunde, R. Wong, J. Wicker, D. Kozlowski, J. Jordan, B. Hansen, G. Muehlnikel, W. Scheel, U. Sterr, R. Kahle and R. Meyer (2009). NFIRE-to-TerraSAR-X laser communication results: satellite pointing, disturbances, and other attributes consistent with successful performance, Proceedings SPIE, vol. 7330, 73300Q.

[61] M. Gregory, F. Heine, H. Kämpfner, R. Lange, K. Saucke, U. Sterr and R. Meyer (2010). Inter-satellite and satellite-ground laser communication links based on homodyne BPSK, Proceedings SPIE, vol. 7587, 75870E.

[62] A. Berk, P. Conforti, R. Kennett, T. Perkins, F. Hawes and J. van den Bosch (2014). MODTRAN6: A major upgrade of the MODTRAN radiative transfer code, Proceedings SPIE, vol. 9088, 90880H.

[63] H. Hemmati (Ed.) (2006). Deep Space Optical Communications, Chapter 3. John Wiley & Sons, New Jersey.

[64] S. Piazzolla (2009). Atmospheric Channel, in *Near-Earth Laser Communications*, H. Hemmati (Ed.), CRC Press, Florida.

[65] T. A. Tsiftsis, H. G. Sandalidis, G. K. Karagiannidis and M. Uysal (2009). Optical wireless links with spatial diversity over strong atmospheric turbulence channels, *IEEE Transactions on Wireless Communications*, vol. 8, pp. 951–957.

[66] X. Zhu and J. M. Kahn (2010). Communication techniques and coding for atmospheric turbulence channels, in *Free-Space Laser Communications*, A. K. Majumdar and J. C. Ricklin (Eds.), Springer, New York.

[67] Z. Sodnik and M. Sans (2012). Extending EDRS to Laser Communication from Space to Ground, International Conference on Space Optical Systems and Applications (ICSOS) 2012, Ajaccio, Corsica, France, October 9–12.

[68] S. Constantine, L. E. Elgin, M. L. Stevens, J. A. Greco, K. Aquino, Daniel D. Alves and B. S. Robinson (2011). Design of a high-speed space modem for the lunar laser communications demonstration, Proceedings SPIE, vol. 7923, 792308.

[69] G. Oppenhauser (1997). Silex program status — A major milestone is reached, *Proceedings SPIE*, vol. 2990, pp. 2–9.

[70] B. Smutny, R. Lange, H. Kämpfner, D. Dallmann, G. Mühlnikel, M. Reinhardt, K. Saucke, U. Sterr and B. Wandernoth (2008). In-orbit verification of optical inter-satellite communication links based on homodyne BPSK, *Proceedings SPIE*, vol. 6877, 687702.

[71] W. R. Leeb and P. J. Winzer (2009). Photodetectors and Receivers, in *Near-Earth Laser Communications*, H. Hemmati (Ed.), CRC Press, Florida.

[72] E. A. Dauler, S. B. Robinson, A. J. Kerman, J. K. W. Yang, M. K. Rosfjord, V. Anant, B. Voronov, G. Gol'tsman and K. K. Berggren (2007). Multi-element superconducting nanowire single-photon detector, *IEEE Transactions on Applied Superconductivity*, vol. 17, pp. 279–284.

[73] M. Shaw, K. Birnbaum, M. Cheng, M. Srinivasan, K. Quirk, J. Kovalik, A. Biswas, A. D. Beyer, F. Marsili, V. Verma, R. P. Mirin, S. W. Nam, J. A. Stern and W. H. Farr (2014). A Receiver for the Lunar Laser Communication Demonstration Using the Optical Communications Telescope Laboratory, CLEO 2014, paper SM4J.2.

[74] B. Laurent, G. Planche and C. Michel (2004). Inter-satellite optical communications: from SILEX to next generation systems, Proceedings of the 5th International Conference on Space Optics (ICSO 2004), March 30 — April 2, Toulouse.

[75] R. Lange and B. Smutny (2007). Homodyne BPSK-based optical inter-satellite communication links, *Proceedings of SPIE*, vol. 6457, 645703.

Chapter 4

Optical Signal Processors and Optical RF Oscillators

In Chapter 3, we described the photonic technology to transfer data or high-purity analog signals between closely spaced units mounted on board a spacecraft, and transmit data from a satellite to the Earth or inter-satellites. In the last four decades, considerable research effort has demonstrated that the same technologies enable more complex functionalities, i.e. the generation of high-frequency analog signals and the processing/routing of both RF signals and data [1]. The result of this effort is the demonstration, in most cases at prototype level, of several photonic devices and sub-systems enabling the implementation of key functionalities, e.g. wideband beamforming or accurate analog-to-digital conversion of millimeter-wave signals. The intrinsic features of these sub-systems, i.e. small size/weight, low power consumption, large tunability, and strong immunity to electromagnetic interference, are very attractive for the space industry, but their fully engineered demonstration is still an R&D target.

Many functionalities of telecom and radar payloads can be realized in the photonic domain, e.g. local oscillator signal generation, frequency conversion, analog-to-digital conversion, and beamforming. However, to the best of our knowledge, photonic technologies for signal generation and processing on board spacecraft are still far from the commercialization stage. In fact, they are nearly all at technology readiness level (TRL) 4. Only one miniaturized opto-electronic oscillator operating at about 35 GHz, also intended for satellite communication systems, is available on the market [2]. However, no data on the influence of the space environment on the features of that component is available in the literature.

One of the first envisaged space applications of optical signal processors is in the field of on-board fast data identification/classification. Some low power miniaturized opto-electronic devices comparing a set of analog voltages, as the outputs of sensing devices, have been conceived and designed with a configurable set of reference signals [3–5]. The device operation aims at avoiding the ground transmission of useless data.

In this chapter, we will discuss how photonics can be used to improve the features and extend the functionalities of several sub-systems for signal generation/processing mounted on board satellites, paying special attention to the technologies that have already demonstrated a high level of compactness and energy/mass saving.

4.1 Opto-electronic oscillators

The generation of high purity sine-wave signals is a key functionality in spacecraft since it is required in all systems that receive and transmit RF signals, including telecom and radar payloads. Oscillators with a low level of phase noise will be required in future telecom payloads due to more stringent requirements related to the use of complex modulation formats and/or the adoption of multiple frequency conversion schemes.

High purity electronic oscillators in the range 10–100 MHz are typically based on piezoelectric quartz resonators, which exhibit a very high Q-factor in that frequency range. Sine-wave signals at higher frequency, i.e. GHz or tens/hundreds GHz, can be obtained using a quartz oscillator with several stages of frequency multipliers and amplifiers at its output. Unfortunately, the multiplication process also implies the multiplication of the noise of the quartz oscillator. Since the phase noise of such oscillators is proportional to N_m^2, with N_m being the frequency multiplication factor, their spectral purity degrades as the frequency increases. In addition, the stages of multipliers and amplifiers increase the mass and the size of the oscillator.

An alternative way to generate a ultra-pure microwave/millimeter-wave signal is the use of the opto-electronic oscillator (OEO) [6], which, starting from the 1990s, has been the topic of intense research effort

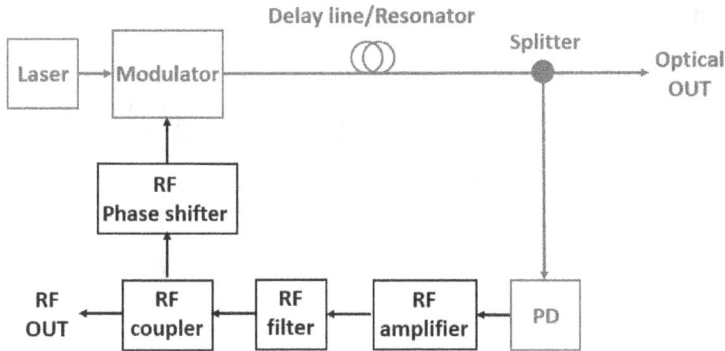

Fig. 4.1. Block diagram of the OEO.

carried out by several research groups [7–9]. The basic configuration of the OEO is shown in Fig. 4.1. The laser generated continuous-wave beam is modulated and then it passes through a long (up to 16 km) optical fiber delay line before reaching a photodetector (PD). The output of the PD is amplified, filtered, phase shifted, and fed back to the electric port of the modulator through an RF coupler. An optical splitter and the RF coupler are used to provide both the optical and the RF outputs.

Assuming that the modulation is performed through a Mach–Zehnder interferometer, the optical power of the beam at the PD input is equal to [10]

$$p(t) = \frac{\alpha P_0}{2}\left\{1 - \sin\left[\frac{\pi V_{in}(t)}{V_\pi} + \frac{\pi V_b}{V_\pi}\right]\right\}, \qquad (4.1)$$

where $\alpha(<1)$ is the modulator insertion loss, $V_{in}(t)$ is the voltage signal driving the modulator, V_b is the bias voltage applied to the modulator, V_π is the half-wave voltage of the modulator, and P_0 is the optical power of the beam generated by the laser source. The voltage at the photodiode output is

$$V_{pd}(t) = R Z P(t), \qquad (4.2)$$

where R is the responsivity of the PD, and Z is the load impedance of the PD. The voltage at the amplifier output is $V_{out} = G_a V_{pd}$ with G_a being the gain of the RF amplifier. The open loop small signal gain of the OEO is equal to [10]

$$G = \frac{\partial V_{out}}{\partial V_{in}}\bigg|_{V_{in}=0} = -\frac{\pi}{V_\pi}\frac{\alpha P_0 R}{2} ZG_a \cos\left(\frac{\pi V_b}{V_\pi}\right). \qquad (4.3)$$

Equation (4.3) shows that the maximum value of $|G|$ is obtained when $V_b = 0$ or $V_b = V_\pi$. The OEO only oscillates if $|G| \geq 1$, thus,

$$\frac{\pi}{V_\pi}\frac{\alpha P_0 R}{2} ZG_a \geq 1. \qquad (4.4)$$

The oscillation condition in Eq. (4.4) implies that to achieve signal generation, the laser power has to be larger than a threshold value P_{th}:

$$P_{th} = \frac{2V_\pi}{\pi \alpha RZG_a}. \qquad (4.5)$$

Typically, P_{th} is of the order of a few tens of mW. This value of laser power is usually obtained by laser diodes or optically pumped solid-state laser sources, e.g. Nd:YAG lasers.

The frequency response of the OEO exhibits equally spaced peaks each related to a resonant mode. The spacing between adjacent peaks is equal to $1/\tau$, where τ is the total group delay of the feedback loop. Due to the presence of the RF filter in the loop, the gain of only one mode can be larger than unity. Thus, the filter selects the oscillating mode and consequently the frequency of the generated sine-wave signal. The amplitude of that signal is approximately equal to [10]

$$V_{osc} \cong \frac{2\sqrt{2}V_\pi}{\pi}\sqrt{1-G^{-1}}. \qquad (4.6)$$

The spectrum of the generated signal exhibits a narrow peak with a full width at half maximum proportional to $1/\tau^2$. The phase noise of the OEO mainly depends on the frequency offset from the oscillation frequency. For a given value of frequency offset, the phase noise is proportional to $1/\tau^2$.

Theoretical and experimental investigations have shown that the phase noise of the OEO is independent of the oscillation frequency [10]. This result is very important because it confirms the OEO potential in generating high frequency and low phase noise signals.

The delay induced by the fiber dominates τ and thus the fiber length L determines the spectral purity of the signal generated by the OEO. The frequency of the generated sine-wave signal is only limited by the bandwidth of the modulator, which is, for example, 40 GHz for a traveling-wave amplitude Mach–Zehnder LiNbO$_3$ modulator to be used in wavelength division multiplexing systems. The required bandwidth of the RF pass-band filter depends on the fiber length L, i.e. the required filter selectivity increases as the fiber length increases. In fact, the OEO is a multimode device, with mode spacing approximately equal to c/nL, c being the speed of light in vacuum and n the fiber refractive index. For example, if the fiber length is 1 km, the required filter bandwidth is of the order of 100 kHz. This kind of performance is unpractical, especially if the frequency signal generated by the OEO is of the order of tens of GHz. An OEO configuration based on an additional fiber loop serving as a filter has been proposed to circumvent this critical aspect [11]. The double loop OEO (see Fig. 4.2) includes two fiber loops of different lengths. The bandwidth required to the pass band filter is determined by the length of the shorter loop, while the oscillator phase noise only depends on the length of the longer loop.

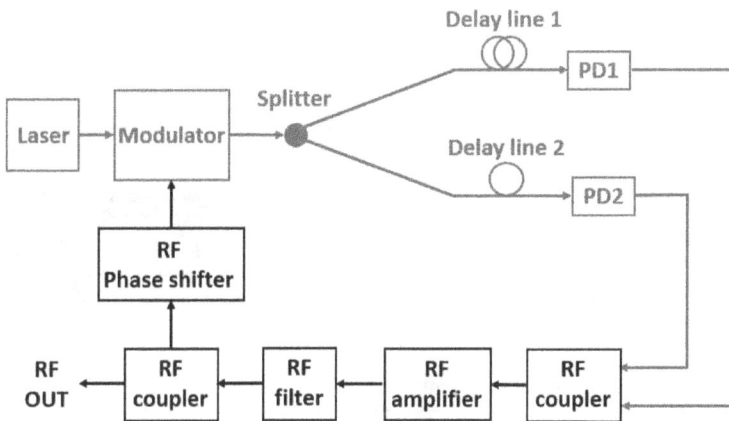

Fig. 4.2.   Block diagram of the double loop OEO.

The OEO configuration discussed until now has demonstrated very good performance, using a Nd:YAG laser with an output power >100 mW, a LiNbO$_3$ traveling-wave modulator, and a fiber length >10 km. In particular, an oscillator operating at 10 GHz based on a thermally stabilized fiber delay line, having a length of 16 km, with phase noise equal to −163 dBc/Hz at 6 kHz offset from the carrier is reported in Ref. 12.

To miniaturize the OEO down to a volume <1 cm^3, an alternative configuration, called coupled opto-electronic oscillator (COEO) has been both theoretically and experimentally investigated [13]. The delay element included in the COEO is either an ultra-high-Q ($Q > 10^8$) optical resonator or a fiber delay line with a length <1 km. The COEO (see Fig. 4.3) includes both an RF section and an optical loop. The optical loop consists of the resonator or the fiber delay line, an optical amplifier and an optical modulator. The RF section includes a PD, an RF amplifier, and an RF filter. To increase the miniaturization level, the modulator functionality can be performed by the resonator, through appropriate tuning. This design choice implies that the optical cavity must be manufactured on an electro-optic substrate. In addition, the filtering function can also be assured by the optical resonator.

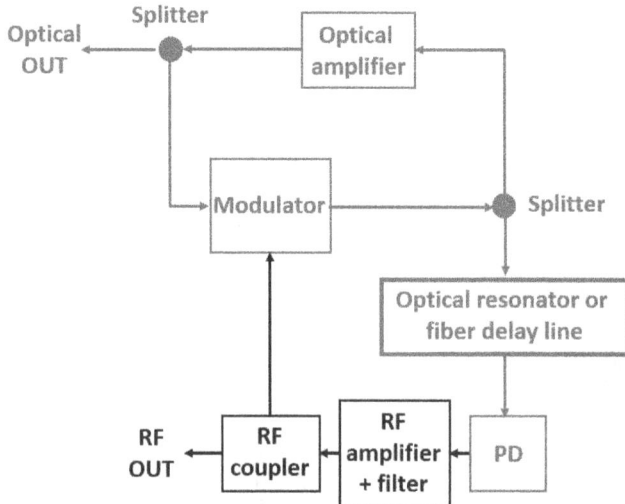

Fig. 4.3.   Block diagram of the COEO.

The key parameter in COEO design is the Q-factor of the optical resonator because it determines the spectral purity of the generated signal. A calcium fluoride crystalline whispering gallery mode resonator, intended for this specific application, having a Q-factor exceeding 10^{11}, is described in Ref. 14. The cavity is an oblate spheroid with a diameter of 4.5 mm and a thickness of 0.5 mm; its optical loss is not limited by surface scattering but only by material attenuation.

A miniaturized COEO generating a sine-wave signal at a fixed frequency in the range 34–36 GHz with a phase noise equal to –108 dBc/Hz at 10 kHz offset from the carrier is currently available in the market [15].

OEO and COEO generate both the RF sine-wave signal and the same signal in optical form. The latter can be used for opto-electronic signal mixing, which will be discussed in the following section.

Alternative approaches for RF sine-wave signal generation through optical techniques are based on optical heterodyning, i.e. the interference of two optical carriers on a photodiode that generates a microwave/ millimeter-wave oscillating signal. For example, by using a dual-frequency laser that generates two optical carriers spaced at 0.25 nm and a large bandwidth photodiode, an RF signal at approximately 30 GHz can be generated from the beating of the two carriers onto the photodiode [16]. The phase noise of the RF signal can be reduced by locking the phase of the two carriers using an optical phase-locked loop.

A quite simple technique for generating phase locked laser lines is based on a fiber mode-locked laser (MLL) with a splitter at its output and two pass band optical filters, centered at different frequencies, at the splitter outputs [17] (see Fig. 4.4). By heterodyning two laser modes from a fiber mode-locked laser, RF sine-wave signals can be generated with a frequency up to a few tens of GHz. The phase noise is in the range from –110 dBc/Hz to –130 dBc/Hz at 10 kHz offset from the carrier [18].

The two carriers at different frequencies interfering on a photodiode can also be generated by double side-band modulation with carrier suppression (see Fig. 4.5) [19].

According to the double side-band modulation with carrier suppression approach, the laser generated continuous-wave beam is amplitude modulated by a Mach–Zehnder modulator, biased at V_π. The modulating RF signal is a sine-wave signal with frequency $f_s/2$, with f_s being the frequency

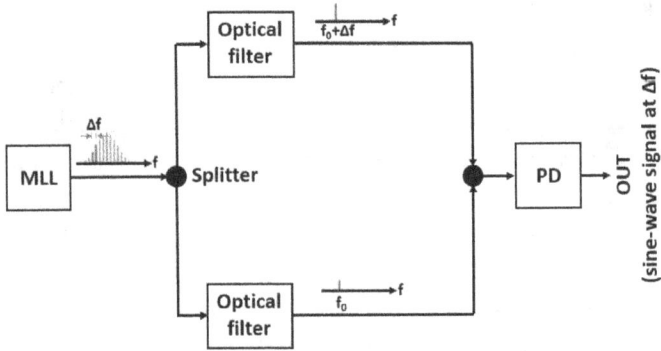

Fig. 4.4.   Optoelectronic oscillator based on a MLL.

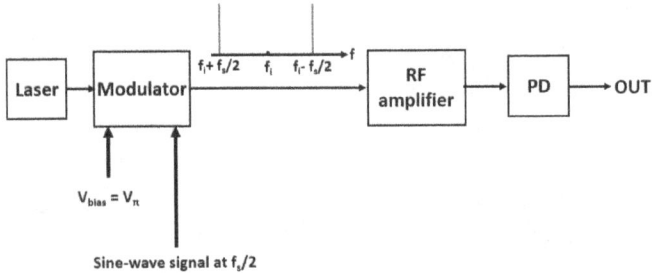

Fig. 4.5.   Opto-electronic oscillator based on double side-band modulation with carrier suppression.

of the RF signal to be generated. If the amplitude of the sine-wave signal at $f_s/2$ is properly adjusted, the spectrum of the beam at the modulator output exhibits only two components at $f_1 + f_s/2$ and $f_1 - f_s/2$ (f_1 is the laser frequency), with the carrier at f_1 being suppressed because $V_b = V_\pi$. The beam at the modulator output is amplified and then sent to the photodiode. The photodiode output is the sine-wave RF signal having frequency f_s.

4.2 Opto-electronic frequency converters

Frequency up- and down-conversion is essential in all telecom payloads, which receive the RF signals and then process and re-transmit them at a frequency ranging from a few GHz to a few tens of GHz. Since the processing is typically performed at low frequency, the received signals have to be

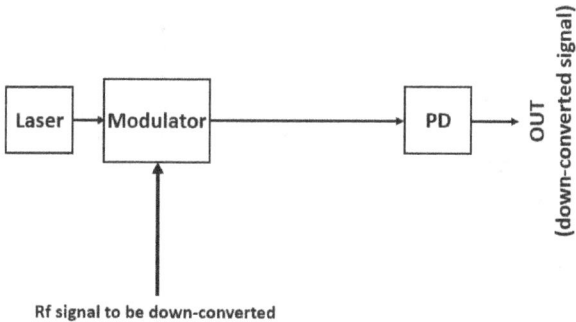

Fig. 4.6.   Opto-electronic down-converter.

down-converted, while the signals to be re-transmitted have to be up-converted. Frequency up- and down-conversion can be performed in the opto-electronic domain using different approaches [20].

The basic principle scheme of an opto-electronic down-converter based on an electro-optic modulator is shown in Fig. 4.6 [21]. The RF signal to be down-converted drives the electro-optic modulator (EOM). The optical beam at the EOM input is the local oscillator (LO) in optical form, i.e. a continuous-wave laser beam modulated by a RF sine-wave signal. The modulator output is O/E transduced by a photodiode. The electric signal at the output of the O/E transduction section is the down-converted RF signal at the intermediate frequency (IF). Opto-electronic microwave mixers have several advantages over their RF counterparts, such as very wide bandwidth of operation, high isolation between the RF and the LO ports, and immunity to electromagnetic interference. The optical form of the LO can be obtained by transducing the LO signal in the RF domain into the optical domain using either a directly modulated laser (if the frequency of the LO is less than a few GHz) or a Continuous Wave (CW) laser and an external EOM. Alternatively, the LO signal in optical form can be provided by an OEO/COEO.

One of the most critical performance parameters of mixers is the conversion efficiency. It is defined as the ratio of the output IF signal power to the input RF signal power. It is quite low in the down-converter shown in Fig. 4.6. The basic approach to solve this critical aspect is the suppression of the optical carrier [22], which implies the increase in the amplitude of

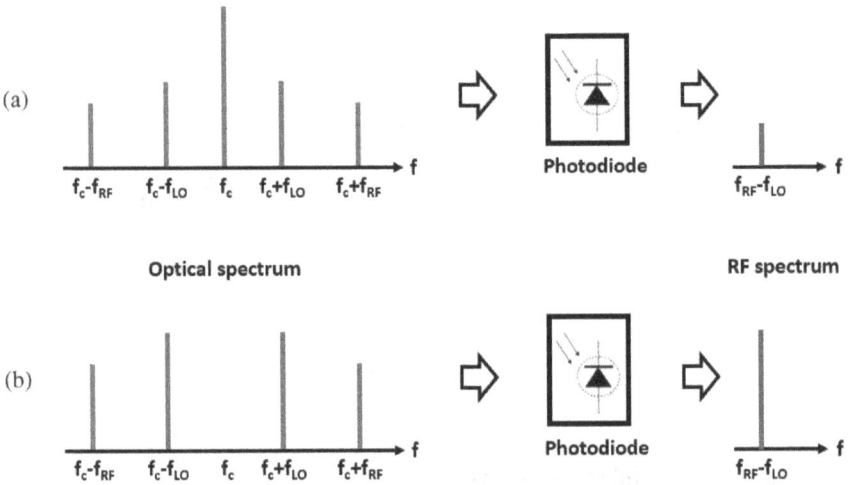

Fig. 4.7.   Opto-electronic down-conversion (a) without carrier suppression and (b) with carrier suppression. f_{LO}: frequency of the local oscillator signal; f_{RF}: frequency of the signal to be down-converted; f_c: frequency of the optical carrier.

the RF signal and LO sidebands in the spectrum of the beam incident on the PD. This results in a higher output IF signal power and consequently higher conversion efficiency. Figure 4.7 schematically illustrates this concept, showing the optical/electrical (O/E) spectra of the signals at the photodiode input/output when carrier suppression is implemented or not.

Several configurations of microwave photonic mixers aimed at improving the conversion efficiency have been investigated in the last few years [23, 24]. Currently the best values of conversion efficiency are of the order of 10 dB.

4.3 Microwave photonic filters

RF filters with fast tunability/reconfigurability and strong selectivity are required for several space applications, e.g. digital telecom payloads, but their realization using conventional technologies, i.e. microstrip or co-planar waveguide lines and substrate-integrated waveguides, is critical. Microwave photonic filters allow this issue to be overcome by exploiting the typical advantages of photonic systems [25].

In microwave photonic filters, the input RF signal is first E/O trans-duced, then the filtering is performed in the optical domain using a

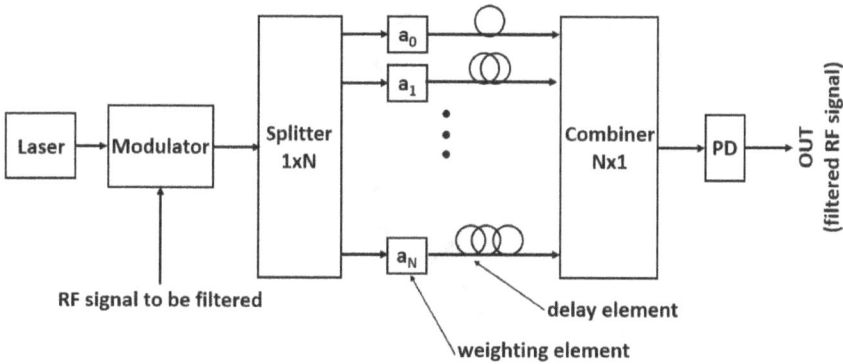

Fig. 4.8. Basic configuration of a microwave photonic filter.

number of delay elements, and finally, the filtered optical beam is O/E transduced. In addition to the tunability/reconfigurability features and very good selectivity, this filtering technology guarantees low loss, high bandwidth, and immunity to electromagnetic interference.

The principle scheme of one of the possible implementations of a microwave photonic filter is shown in Fig. 4.8. The RF signal to be filtered drives an EOM, which modulates an optical carrier generated by a continuous wave laser source. The optical beam at the modulator output is split into several replicas that are weighted and delayed in the optical domain. Finally, the replicas are combined and the resulting beam is transduced in the RF domain by a photodiode. The interference of all the delayed and weighted modulated optical signals in the combiner enables the filtering functionality.

Recently, the technology of microwave photonic filters has led to the experimental demonstration of a tunable bandpass filter with a narrow flat-top shape operating in the Ka-band, which is of interest for telecom payloads. The filter has a 3 dB bandwidth of 840 MHz [26].

4.4 Photonic switching

Switching is a core function for both digital and analogue telecom payloads [27]. The technology of micro-opto-electro-mechanical systems (MOEMS) switches for space applications has been recently assessed and developed. MOEMS switches route the optical beam from one fiber to

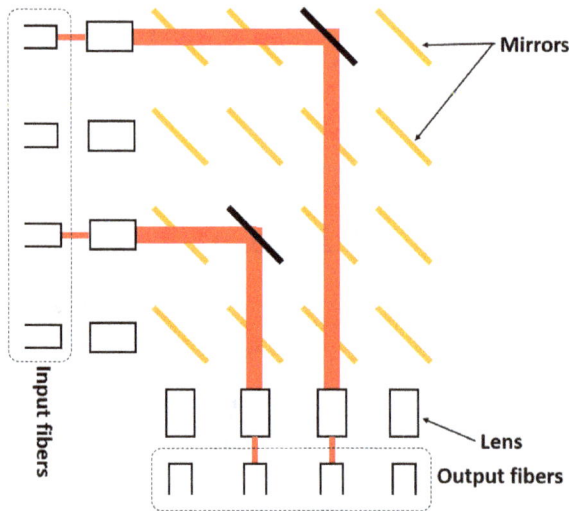

Fig. 4.9.   2D MOEMS switch.

another by steering the light through a collimating lens, reflecting it off a movable mirror and redirecting it into one of the output ports [28, 29].

Two configurations, based on either a 2D array of mirrors or a 3D structure, have been investigated. In the 2D structure, (see Fig. 4.9) the micromirrors and the fibers are arranged in a planar configuration and N^2 mirrors are used to connect N input fibers to N output ones. The mirrors can only be in two positions at any given time. In the 3D approach, two arrays of N mirrors are used to connect N input fibers to N output fibers. Each mirror can assume at least N different positions. The manufactured 2D switch supports up to 10 input/output ports while the 3D configuration allows an increase by a factor 5 in the number of supported input/output ports.

4.5 Photonic analog-to-digital converters

The digital processing capabilities, that are required for telecom payloads, are quickly growing and thus, the R&D effort on on-board digital processors is currently increasing. Analog-to-digital conversion, providing data to these processing sub-systems, is currently performed after the down-conversion of the RF beams received by the satellite. The mass and power consumption of the RF down-conversion sub-system are very large if the number of channels received, routed, and re-transmitted by the telecom satellite exceeds a few hundreds. For example, the RF down-conversion sub-system has a mass of approximately 180 kg and a power consumption of 1.6 kW if the number of channels is 200 [30]. The direct analog-to-digital conversion of the received RF beams, without any down-conversion, could overcome this issue, but the currently available space-qualified electronic analog-to-digital converters (ADCs) allow the direct conversion of RF signals up to a few GHz [31]. Since the RF signal received by the future telecom satellites will be typically in the Ka-band (30 GHz uplink and 20 GHz downlink), photonic ADCs can be very useful to overcome the limitations of electronic ADCs. In addition, the features of photonic ADCs can be very useful in the context of next-generation radar systems, which will exhibit very high carrier frequencies and very broad bandwidth to enable the use of smaller antennas and give higher resolution with respect to the state-of-the-art [32].

Photonic ADCs, which have been widely investigated in the last four decades, exploit photonic devices to digitalize an analog RF signal and generate a digital electronic output [33]. Photonic ADCs, where sampling and quantization are both performed in the electronic domain, and photonics is used only to improve one of these functionalities, e.g. sampling, are called photonic assisted ADCs. If either sampling or quantization are performed in the optical domain, the ADCs are called photonic sampled or photonic quantized, respectively. Finally, ADCs where both the basic functionalities, i.e. sampling and quantization, take place in the photonic domain are called photonic sampled and quantized.

Photonic sampled ADCs, which are the most mature and promising among the photonic ADCs, allow the limitations imposed to electronic ADCs by aperture jitter, i.e. the temporal variation of the exact sampling time instant, to be overcome. In particular, aperture jitter limits the effective number of bits (ENOB) [34] of the ADC, which quantifies the noise and the distortion introduced by the converter, and increases as the frequency of the input analog signal increases. Therefore, the ADC ENOB decreases as the analog input frequency decreases. Currently, due to the physical limitations imposed by electronic devices, the jitter of the best electronic ADC is approximately 100 fs, thus electronic ADCs exhibit an

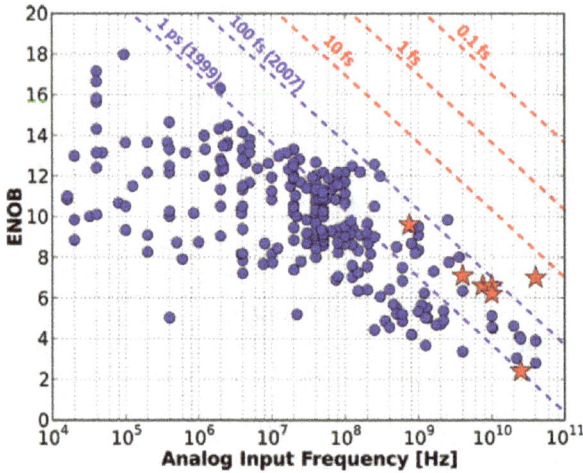

Fig. 4.10. ENOB *versus* analog input frequency for electronic and photonic ADCs. Reprinted from Ref. [35], with the permission of IEEE.

ENOB <6 when the input frequency is ≥10 GHz. The use of photonic technology allows jitter values close to 10 fs with an improvement of about one order of magnitude with respect to electronic technology.

ENOB dependence on the input frequency for both electronic (dots) and photonic (stars) ADC is shown in Fig. 4.10. Dashed lines represent the limitation imposed by the jitter on the ADC performance.

The best results in terms of ENOB at high frequency have been achieved by two devices reported in Refs. [36, 37]. Both the photonic sampled ADCs, whose operating principle is schematically shown in Fig. 4.11, exhibit an ENOB of 7 when the analog input frequency is approximately 40 GHz. In both the devices, narrow-pulse mode-locked lasers are used for sampling, which is accomplished when a pulse train generated by the laser passes through an electro-optic modulator driven by the RF electric signal to be A/D converted. At the modulator output, the energy associated with each pulse is proportional to the amplitude of the RF signal at the temporal positions of the pulses. The advantage of this approach is that ADC jitter is determined by the jitter of the optical pulse train, which is typically very low (of the order of 10 fs). At the modulator output, the pulse train can be split into multiple parallel pulse trains having a lower repetition rate using various techniques in the time (Fig. 4.11(a))

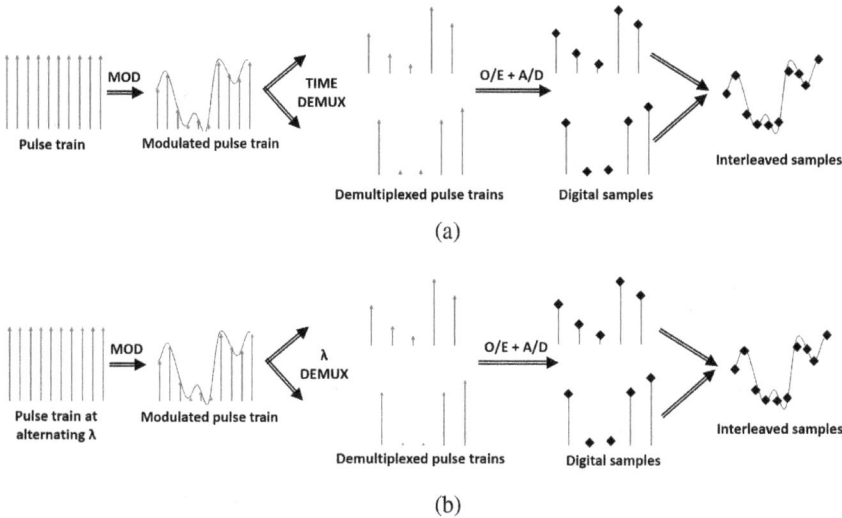

Fig. 4.11. Operating principle of photonic sampled ADCs.

or wavelength domain (Fig. 4.11(b)). These pulse trains are then O/E transduced by large bandwidth photodiodes. The photodiode outputs are finally A/D converted by electronic ADCs and the digital samples are time-interleaved. Some preliminary results on a photonic sampled ADC for space applications have been reported in Ref. [30].

Recently, a chip implementing some of the key functionalities of a photonic sampled ADC, i.e. modulation, wavelength demultiplexing, and photodetection has been fabricated in silicon photonics. The device exhibits an ENOB of 3.5 with an analog input frequency of 10 GHz [38].

4.6 Photonic beamformers

A phased-array antenna (PAA) [39] consists of an array of multiple stationary antenna elements, which are closely spaced. The distance between the elements is generally of the same order as the wavelength of the RF signal to be transmitted. The direction and the characteristics of the transmitted beam are tuned by controlling the amplitude and the phase of the RF signal feeding each antenna element and the overall radiation pattern results from a combination, in amplitude and phase, of the waves radiated by the antenna elements. The most important advantage of PAAs is their ability to produce a very precisely directed beam, which can be quickly scanned electronically. This feature of PAAs is extremely attractive for several space systems, such as telecom or radar payloads, with a consequent very wide use of PAAs in space engineering. For example, all synthetic aperture radar (SAR) satellites, which have been launched in recent years, e.g. Radarsat-2, the four satellites forming the Cosmo Skymed constellation, TerraSAR-X, and TanDEM-X, utilize PAA technology [40–42]. The advantages of PAAs also include their ability to generate multiple simultaneous antenna beams, as required by multi-beam telecom payloads.

In passive PAAs (see Fig. 4.12), a central high-power RF source is used to generate the signal, which is divided by a power splitter into many signals, feeding the antenna elements after their attenuation and phase shifting. The attenuation and the phase shift imposed on each signal is electronically controlled. The amplitude of each signal determines the shape of the transmitted beam (beam shaping), while the direction of the transmitted beam is set by regulating the phase of each signal (beam steering). Beam shaping and beam steering enable so-called beam forming. Circuits performing the beamforming are called beamformers.

In active PAAs (see Fig. 4.13), the central high-power RF transmitter is replaced by solid-state RF amplifiers directly connected to each antenna element. In addition to active and passive PAAs, there is an intermediate

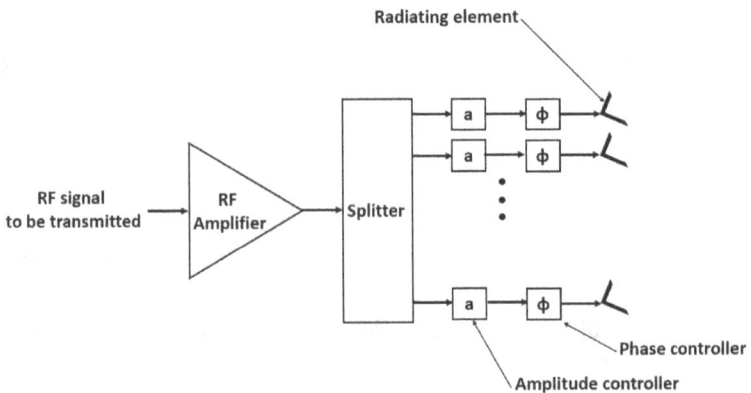

Fig. 4.12.   Block diagram of a passive PAA.

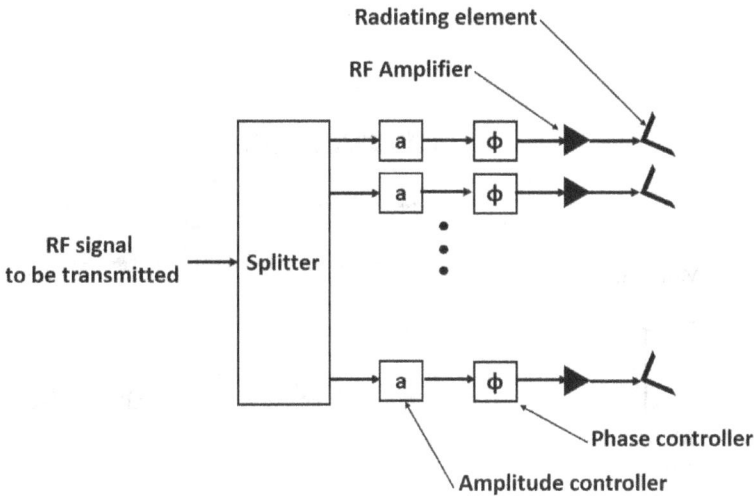

Fig. 4.13.   Block diagram of an active PAA.

solution, consisting of dividing the radiating elements into groups called sub-arrays. Each sub-array is fed by an RF amplifier.

The phase shift imposed by an RF phase shifter depends on the frequency of the input signal. Thus, for a wideband RF signal, the different spectral components suffer from a non-uniform phase shift. In PAAs, this critical aspect generates so-called beam squinting, i.e. a lack of uniformity in beam pointing for the different spectral components of a wideband RF signal. To circumvent this issue, true-time-delay phase shifters are utilized, imposing a phase shift that is frequency-independent in a large frequency range.

When the PAA operates in the Ka- or Q-band (frequency >20 GHz), the generation of true time delays in the RF domain can become critical because bulky, leaky, and heavy circuits are required. Therefore, the use of photonics to generate true time delays for PAAs has been the topic of extensive research effort in the last two decades [43]. Currently, photonic beamformers have already demonstrated with broad instantaneous bandwidth, continuous amplitude and delay tunability as well as capability of feeding large arrays.

Although free-space beamformers based on spatial light modulators [44, 45] having a very high degree of parallelism, i.e. able to control thousands of antenna elements simultaneously, have been widely investigated,

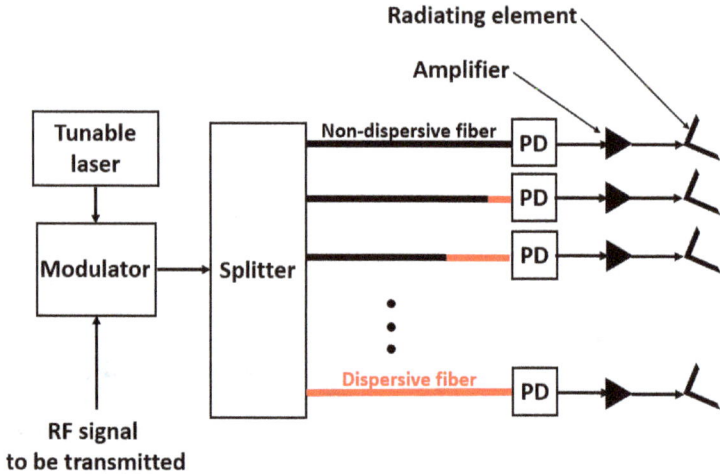

Fig. 4.14.   Beamformer based on the fiber optic prism.

current research trends seem to be focused on beamformer miniaturization using optical fibers or integrated optics.

The first beamformers based on fiber optic delay elements were demonstrated at the end of the 1980s [46, 47]. For example, a beamformer based on the so-called fiber optic prism (see Fig. 4.14) was reported in Ref. [48]. The beam generated by a tunable laser source is amplitude modulated by the microwave signal to be transmitted. The modulated optical beam is then split and the outputs of the splitter propagate in several fiber links with slightly different dispersion. At the end of each link, O/E transduction takes place and the RF signals feeding the antenna elements are obtained. Since the fiber links have the same length but different dispersion, the time delay imposed by each of them is different. In addition, the delay suffered by the beams propagating in the different links depends on the wavelength of the laser beam. Thus, the time delay between the beams at the outputs of the fiber links are tuned by varying the wavelength of laser source. A prototype of an optically-steered X-band 8-element PAA based on this approach, with capabilities of multi-band and multi-beam operations, has been recently demonstrated [49]. The system volume is 7.6 cm × 20.3 cm × 43.2 cm and its beam steering speed is 15°/μs.

The fiber optic prism concept has also been implemented by using fiber Bragg gratings and photonic crystal (PhC) waveguide arrays [50, 51].

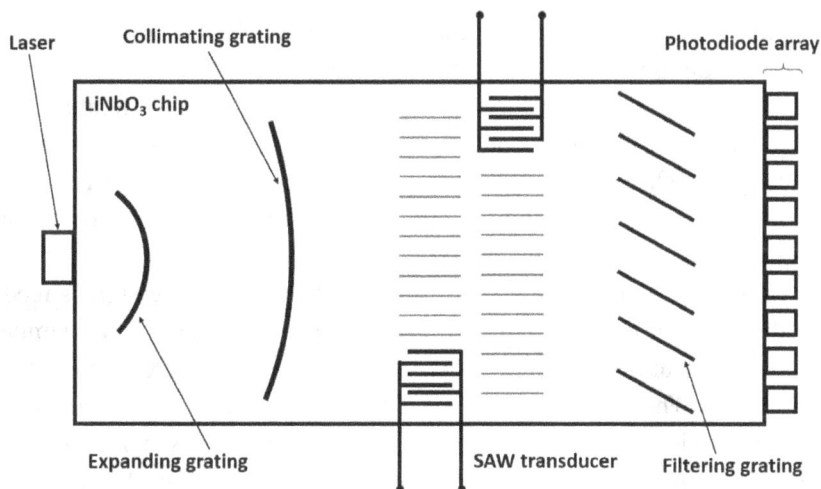

Fig. 4.15.   LiNbO$_3$ integrated optical beamformer based on two SAW transducers.

Aiming to reduce the beamformer size and increasing the accuracy of the time delay imposed onto the beams feeding the antenna elements, several configurations of integrated optical beamformers have been conceived, designed, and fabricated.

The design of one of the first LiNbO$_3$ integrated optical beamformers based on two surface acoustic-wave (SAW) transducers is reported in Refs. [52, 53] (see Fig. 4.15).

The laser beam, which is modulated by the RF signal to be transmitted by the PAA, is coupled into the device, where it is expanded and collimated by two nonlinear grating lenses. The optical beam coming out from the lens interacts with the two counter-propagating SAWs. The acousto-optic grating, due to the SAW, generates only two output orders for each incident mode, one diffracted and the other transmitted. Thus, after the interaction with the SAWs, four modes come into the filtering grating. Only the two diffracted modes pass through the filter and interfere on the array of detectors. The phase of the RF signals at the output of each detector depends on the detector position in the array and the difference between the frequencies of the electric signals driving the SAW transducers. The tuning of that difference results in very effective beam steering. The designed device footprint is 40 mm × 15 mm, its phase error is <2°, and the maximum position beam error is approximately 0.5°.

The integrated optical beamformers reported in the literature are either wideband or narrowband. Narrowband beamformers are based on optical phase shifters and coherent detection while wideband beamformers, which can be discretely or continuously tunable, are all based on photonic true-time-delay elements.

In narrowband photonic beamformers, the phase of the signals feeding the antenna element is controlled in the optical domain by simple compact optical phase shifters exploiting the thermo-optic or the electro-optic effect. The phase of the optical beam is directly transferred to the microwave signal by an appropriate coherent detection scheme. Several photonic beamformers based on this operating principle have been experimentally demonstrated [54–56]. In particular, a beamformer intended for multi-beam telecom payloads with a standard deviation of phase errors <5° has been proposed [57].

Several wideband beamformers, which are intended for radio astronomy and reception of Ku-band RF signal transmitted by satellites for digital video broadcast, have been recently demonstrated [58–60]. The beamformers use single or cascaded optical ring resonators as a tunable delay element. The tuning is achieved by a phase shifter included in the resonant path and/or controlling the coupling between the ring and the bus waveguide. The rings are arranged according to a binary tree topology, which implies the use of several tunable combiners. The block diagram of the PAA system for satellite signal reception, including an integrated optical beamformer is shown in Fig. 4.16(a). The RF signals coming out from the PAA elements are low-noise amplified, frequency down-converted, and E/O transduced using one laser with a splitter at its output and an electro-optic modulator for each channel. The optical beams at the modulator outputs are properly delayed and combined in the beamformer. O/E transduction is performed by a photodiode, which generates the received RF signal. The beamformer chip (Fig. 4.16(b)) includes both single and cascaded optical ring resonators, and tunable combiners. A beamformer based on this technology, having a footprint of 3.6 cm × 0.8 cm and an instantaneous bandwidth of 8 GHz has been very recently demonstrated [61].

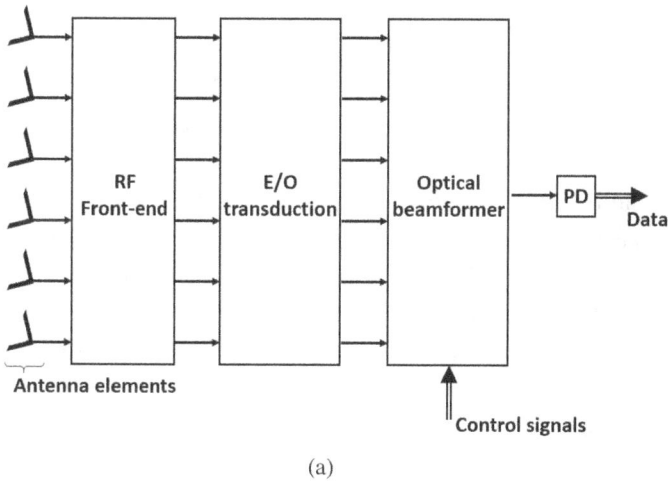

(a)

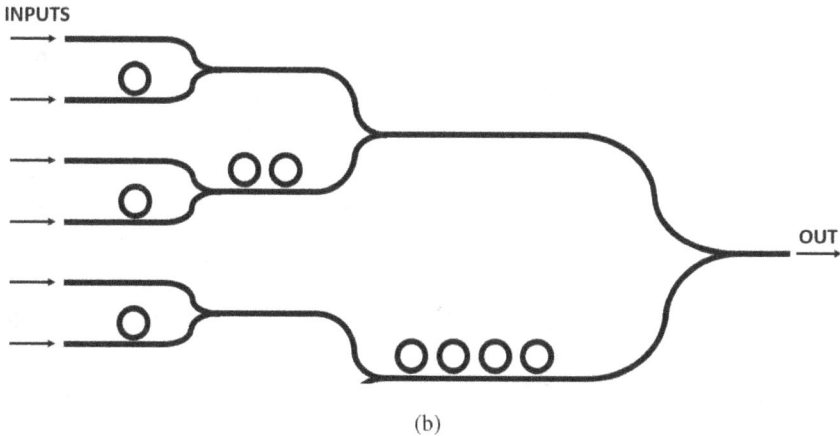

(b)

Fig. 4.16. (a) Block diagram of the PAA system for satellite signals reception. (b) Beamformer chip including single and cascaded optical ring resonators.

4.7 Opto-electronic SAR processors

Since the NASA space mission SeaSat [62] (launched on June 27, 1978) utilizing a synthetic aperture radar (SAR) for remote sensing of the Earth's oceans, SAR payloads have been unanimously considered a very powerful tool for the Earth observation because they exhibit a very good spatial resolution and can operate in all weather conditions both during day and at night.

The SAR [63] was developed in the 1950s with the aim of achieving broad-area radar imaging at high resolution and it allows an extremely large antenna to be simulated by utilizing the relative motion between the antenna and its target region. Therefore, SAR systems are mounted on board spacecraft and illuminate the target scene with a periodic pulse train as the spacecraft moves. The sequence of radar echoes coming from the target are coherently detected by the radar instrument at different locations along the spacecraft's path. The sequence of several radar echoes is processed to generate the SAR image.

The typical SAR imaging scenario is shown in Fig. 4.17. The axes of the SAR image are the range and the azimuth. The former is defined as the direction from the SAR payload to the center of the target scene, while the latter is the direction along the satellite track. The two basic operations of SAR processing are range compression and azimuth compression, which are basically two correlation operations. The width of the ground area covered during a single pass of the satellite is called the swath width.

In the first SAR systems, bulk units based on Fourier optics were utilized for ground processing of radar echoes, which were often stored on photographic film. Optical ground processing of the film recorded SAR data that was widely utilized until the 1980s. In recent decades, ground processing of SAR data has been performed digitally, using several

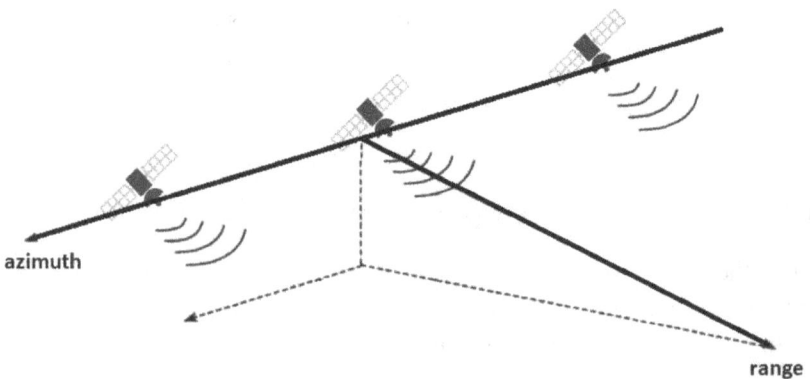

Fig. 4.17.   Typical SAR imaging scenario.

algorithms based on 2D fast Fourier transform (FFT). This processing operation is usually performed in large calculation infrastructures.

It is well known that SAR antennas usually generate large amounts of complex-valued raw data. Thus, real-time on-board processing of SAR data by small and lightweight units would be extremely useful, for example in interplanetary missions, where the bandwidth of the transmission channel allowing data transfer to the ground makes the generation of SAR images of large areas with very high resolution very complex. In this context, SAR image formation on board spacecraft may be very useful especially because SAR images can be compressed more easily that SAR row data, i.e. the compression factor increases by about one order of magnitude. Thus, the on-board formation of SAR images could also allow SAR imaging at very high resolution also in the framework of interplanetary missions.

Pioneering research activity on photonic miniaturized systems for SAR radar echo processing is reported in Refs. [64, 65]. The design of a microsystem for SAR on-board processing including a directly modulated laser diode emitting at 0.85 μm, a GaAs chip with a footprint of about 4 cm^2 performing range compression, and a 2D charge-coupled device (CCD) array, which carries out azimuth compression, is presented in Ref. [66]. The configuration of the SAR processor is shown in Fig. 4.18. The laser beam, after being expanded and collimated by two gratings with nonlinear profiles, is deflected by an acousto-optic Bragg cell. This interaction generates a diffracted and an undiffracted beam. The latter is filtered by the filtering grating, while the diffracted beam is coupled into a stripe waveguide array by a Fresnel lens array. The array function is to out-couple the incident beam, which is directed towards the CCD matrix through a transmission mask located in close proximity.

The laser diode is modulated by the transmitted pulse train and the received radar echos are applied to the acousto-optic transducer. The beam diffracted by the Bragg cell has a space-time modulation proportional to the product between the transmitted pulse train and the delayed radar echo. By integrating this product, the correlation of the two signals is obtained and thus, the range compression is performed. This integration is carried out by the CCD array.

Recently, a new prototype of an opto-electronic on-board SAR processor has been developed [67, 68]. It allows the fast generation (in few seconds) of high resolution SAR images with a relatively low power consumption

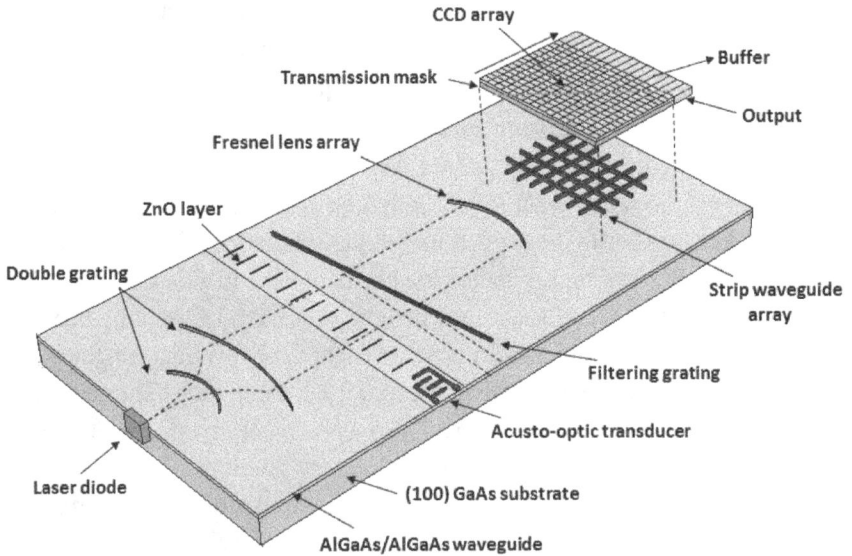

Fig. 4.18.   Configuration of the SAR processor in Ref. [66].

Fig. 4.19.   Prototype of the opto-electronic on-board SAR processor in Ref. [68].
Source: Reprinted from Ref. [68], with the permission of IEEE.

(17 W). The unit, shown in Fig. 4.19, includes a laser diode, two high definition (1920 × 1080 pixel) spatial light modulators (SLMs), several lenses, and
a high-resolution complementary metal–oxide semiconductor (CMOS) camera. SLMs encode complex SAR raw data in amplitude and phase, the lenses

perform the range/azimuth compression, and finally the CMOS camera captures the SAR image. SAR data generated during the Advanced Synthetic Aperture Radar (ASAR) mission [69] was processed in the laboratory by a prototype opto-electronic SAR processor, demonstrating the high quality of the generated images with a processor volume of approximately 17,000 cm^3.

References

[1] D. Marpaung, C. Roeloffzen, R. Heideman, A. Leinse, S. Sales and J. L. Capmany (2013). Integrated microwave photonics, *Laser & Photonics Reviews*, vol. 7, pp. 506–538.

[2] L. Maleki (2011). The optoelectronic oscillator, *Nature Photonics*, vol. 5, pp. 728–730.

[3] M. N. Armenise and V. M. N. Passaro (1989). On-board optical preprocessor on lithium niobate waveguide, *Bulgarian Journal of Physics*, vol. 16, pp. 563–569.

[4] M. N. Armenise and V. M. N. Passaro (1990). A new configuration of an on-board optical preprocessor on lithium niobate waveguide, *Proceedings SPIE*, vol. 1177, p. 123.

[5] M. N. Armenise and V. M. N. Passaro (1990). Design and simulation of an on-board lithium niobate integrated optical preprocessor, *IEE Proceedings Journal (Optoelectronics)*, vol. 137, pp. 347–356.

[6] X. S. Yao and L. Maleki (1994). High frequency optical subcarrier generator, *Electronics Letters*, vol. 30, pp. 1525–1526.

[7] K. Saleh, P.-H. Merrer, A. Ali-Slimane, O. Llopis and G. Cibiel (2013). Study of the noise processes in microwave oscillators based on passive optical resonators, *International Journal of Microwave and Wireless Technologies*, vol. 5, pp. 371–380.

[8] J. Xiong, R. Wang, T. Fang, T. Pu, D. Chen, L. Lu, P. Xiang, J. Zheng and J. Zhao (2013). Low-cost and wideband frequency tunable optoelectronic oscillator based on a directly modulated distributed feedback semiconductor laser, *Optics Letters*, vol. 38, pp. 4128–4130.

[9] X. Xie, C. Zhang, T. Sun, P. Guo, X. Zhu, L. Zhu, W. Hu and Z. Chen (2013). Wideband tunable optoelectronic oscillator based on a phase modulator and a tunable optical filter, *Optics Letters*, vol. 38, pp. 655–657.

[10] X. S. Yao and L. Maleki (1996). Optoelectronic oscillator for pholtonic svstems, *IEEE Journal of Quantum Electronics*, vol. 32, pp. 1141–1149.

[11] X. S. Yao and L. Maleki (2000). Multiloop optoelectronic oscillator, *IEEE Journal of Quantum Electronics*, vol. 36, pp. 79–84.

[12] D. Eliyahu, D. Seidel and L. Maleki (2008). Phase noise of a high performance OEO and an ultra low noise floor cross-correlation microwave photonic homodyne system, *IEEE International Frequency Control Symposium*, Honolulu, HI, USA, May 19–21.

[13] X. S. Yao, L. Davis and L. Maleki (2000). Coupled optoelectronic oscillators for generating both RF signal and optical pulses, *Journal of Lightwave Technology*, vol. 18, pp. 73–78.

[14] A. A. Savchenkov, A. B. Matsko, V. S. Ilchenko and L. Maleki (2007). Optical resonators with ten million finesse, *Optics Express*, vol. 15, pp. 6768–6773.

[15] L. Maleki (2011). The optoelectronic oscillator, *Nature Photonics*, vol. 5, pp. 728–730.

[16] B. Bénazet, M. Sotom, M. Maignan and J. Berthon (2004). Optical distribution of local oscillators in future telecommunication satellite payloads, 5th *International Conference on Space Optics* (*ICSO 2004*), March 30–April 2, Toulouse, France.

[17] P. Ghelfi, G. Serafino, F. Berizzi and A. Bogoni (2010). Generation of Highly Stable Microwave Signals Based on Regenerative Fiber Mode Locking Laser, Conference on Lasers and Electro-Optics (CLEO), San Jose, CA, May 16–21.

[18] G. Serafino, P. Ghelfi, P. Pérez-Millán, G. E. Villanueva, J. Palací, J. L. Cruz and A. Bogoni (2011). Phase and amplitude stability of EHF band radar carriers generated from an active mode-locked laser, *Journal of Lightwave Technology*, vol. 29, pp. 3551–3559.

[19] B. Bénazet, M. Sotom, M. Maignan and J. M. Perdigues Armengol (2005). Optical Technologies for On-Board Processing of Microwave Signals, Potentially Disruptive Technologies and Their Impact in Space Programs, Marseille, France, July 4–6 July.

[20] S. A. Pappert, R. Helkey and R. T. Logan (2002). Photonic link techniques for microwave frequency conversion, in *RF Photonic Technology in Optical Fiber Links*, W. S. C. Chang (Ed.), Cambridge University Press.

[21] G. K. Gopalakrishnan, W. K. Burns and C. H. Bulmer (1993). Microwave-optical mixing in $LiNbO_3$ modulators, *IEEE Transactions on Microwave Theory and Techniques*, vol. 41, pp. 2383–2391.

[22] R. A. Minasian, E. H. W. Chan and X. Yi (2013). Microwave photonic signal processing, *Optics Express*, vol. 21, pp. 22918–22936.

[23] A. Altaqui, E. H. W. Chan and R. A. Minasian (2014). Microwave photonic mixer with high spurious-free dynamic range, *Applied Optics*, vol. 53, pp. 3687–3695.

[24] E. H. W. Chan (2014). Microwave photonic mixer based on a single bidirectional Mach–Zehnder modulator, *Applied Optics*, vol. 53, pp. 1306–1314.

[25] J. Capmany, B. Ortega and D. Pastor (2006). A tutorial on microwave photonic filters, *Journal of Lightwave Technology*, vol. 24, pp. 201–229.

[26] Y. Deng, M. Li, N. Huang and N. Zhu (2014). Ka-Band tunable flat-top microwave photonic filter using a multi-phase-shifted fiber Bragg grating, *IEEE Photonics Journal*, vol. 6, pp. 5500908–5500908.

[27] N. Karafolas, J. M. P. Armengol and I. Mckenzie (2009). Introducing photonics in spacecraft engineering: ESA's strategic approach, IEEE Aerospace conference, Big Sky, MT, USA, March 7–14.

[28] P. Herbst, C. Marxer, M. Sotom, C. Voland, M. Zickar, W. Noell and N. de Rooij (2005). Micro-optical switches for future telecommunication payloads: Achievements of the SAT 'N LIGHT Project, *5th ESA Round Table on Micro/Nano Technologies for Space*, ESA/ESTEC, Noordwijk, The Netherlands, October 3–5.

[29] M. Gobet, P. Herbst, P. Baroni and C. Marxer (2010). Applications of optical MEMS for space missions, *7th ESA Round Table on Micro/Nano Technologies for Space*, ESA/ESTEC, Noordwijk, The Netherlands, September 13–17.

[30] S. Pantoja, M. A. Piqueras, P. Villalba, B. Martínez and E. Rico (2010). High-performace photonic ADC for Space applications, *International Conference on Space Optics*, Rhodes, Greece, October 4–8, 2010.

[31] N. Chantier, B. Dervaux, C. Lambert, V. Monier, C. Allene and O. Bonnet (2012). Challenges of Mixed Signal Space grade ICs operating at Microwave frequencies, AMICSA 2012, ESA/ESTEC, Noordwijk, The Netherlands, August 26–28.

[32] P. Ghelfi, F. Laghezza, F. Scotti, G. Serafino, A. Capria, S. Pinna, D. Onori, C. Porzi, M. Scaffardi, A. Malacarne, V. Vercesi, E. Lazzeri, F. Berizzi and A. Bogoni (2014). A fully photonics-based coherent radar system, *Nature*, vol. 507, pp. 341–345.

[33] G. C. Valley (2007). Photonic analog-to-digital converters, *Optics Express*, vol. 15, pp. 1955–1982.

[34] IEEE (2011). IEEE Standard for Terminology and Test Methods for Analog-to-Digital Converters, IEEE Std 1241-2010.

[35] C. Laperle and M. O'Sullivan (2014). Advances in high-speed DACs, ADCs, and DSP for optical coherent transceivers, *Journal of Lightwave Technology*, vol. 32, pp. 629–643.

[36] A. Khilo *et al.* (2012). Photonic ADC: Overcoming the bottleneck of electronic jitter, *Optics Express*, vol. 20, pp. 4454–4469.

[37] F. Laghezza, F. Scotti, P. Ghelfi, A. Bogoni and S. Pinna (2013). Jitter-limited photonic analog-to-digital converter with 7 effective bits for wideband radar applications, *IEEE Radar Conference*, Ottawa, ON, Canada, April 29–May 3.

[38] M. E. Grein, S. Spector, A. Khilo, A. H. Nejadmalayeri, M. Y. Sander, M. Peng, J. Wang, C. M. Sorace, M. W. Geis, M. M. Willis, D. M. Lennon, T. Lyszczarz, E. P. Ippen and F. X. Kaertner (2011). Demonstration of a 10 GHz CMOS-Compatible Integrated Photonic Analog-to-Digital Converter, CLEO 2011, paper CThI1.

[39] R. J. Mailloux (2005). *Phased Array Antenna Handbook.* Artech House, Norwood.

[40] J. Uher, C. Grenier and G. Lefebvre (2004). RADARSAT-2 SAR antenna, *Canadian Journal of Remote Sensing*, vol. 30, pp. 287–294. A. Torre and P. Capece (2011). COSMO-SkyMed: The advanced SAR instrument, *5th International Conference on Recent Advances in Space Technologies (RAST)*, Istanbul, June 9–11.

[41] W. Pitz and D. Miller (2010). The TerraSAR-X Satellite, *IEEE Transactions on Geoscience and Remote Sensing*, vol. 48, pp. 615–622.

[42] M. Bartusch, H. J. Berg and O. Siebertz (2008). The TanDEM-X Mission, *7th European Conference on Synthetic Aperture Radar (EUSAR)*, Friedrichshafen, Germany, June 2–5.

[43] A. J. Seeds (2002). Microwave photonics, *IEEE Transactions on Microwave Theory and Techniques*, vol. 50, pp. 877–887.

[44] D. Dolfi, F. Michel-Gabriel, S. Bann and J. P. Huignard (1991). Two-dimensional optical architecture for time-delay beam forming in a phased-array antenna, *Optical Letters*, vol. 16, pp. 255–257.

[45] T. Akiyama, H. Matsuzawa, E. Haraguchi, H. Sumiyoshi, T. Ando, A. Akaishi, T. Takahashi, Y. Fujino and R. Suzuki (2011). Spatial light modulator based optically controlled beamformer for variable multiple-spot beam antenna, *International Topical Meeting on Microwave Photonics*, Singapore, October 18–21.

[46] W. Ng, A. Walston, G. Tangonan, I. Newberg and J. J. Lee (1989). Wideband fibre-optic delay network for phased array antenna steering, *Electronics Letters*, vol. 25, pp. 1456–1457.

[47] W. Ng, A. A. Walston, G. L. Tangonan, J. J. Lee and I. L. Newberg (1991). The first demonstration of an optically steered microwave phased array antenna using true-time-delay, *Journal of Lightwave Technology*, vol. 9, pp. 1124–1131.

[48] R. D. Esman, M. Y. Frankel, J. L. Dexter, L. Goldberg, M. G. Parent, D. Stilwell and D. G. Cooper (1993). Fiber-optic prism true time-delay antenna feed, *IEEE Photonics Technical Letters*, vol. 5, pp. 1347–1349.

[49] P. Wu, S. Tang and D. E. Raible (2013). A prototype high-speed optically-steered X-band phased array antenna, *Optics Express*, vol. 21, pp. 32599–32604.

[50] H. Zmuda, R. A. Soref, P. Payson, S. Johns and E. N. Toughlian (1997). Photonic beamformer for phased array antennas using a fiber grating prism, *IEEE Photonics Technical Letters*, vol. 9, pp. 241–243.

[51] C.-Y. Lin, H. Subbaraman, A. Hosseini, A. X. Wang, L. Zhu and R. T. Chen (2012). Silicon nanomembrane based photonic crystal waveguide array for wavelength-tunable true-time-delay lines, *Applied Physics Letters*, vol. 101, 051101.

[52] M. N. Armenise, V. M. N. Passaro, T. Conese and A. M. Matteo (1993). New guided-wave optical beam former for phased array antennas, Proceedings SPIE, vol. 1794, 489.

[53] M. N. Armenise, V. M. N. Passaro and G. Noviello (1994). Lithium niobate guided-wave beam former for steering phased-array antennas, *Applied Optics*, vol. 33, pp. 6194–6209.

[54] K. Horikawa, Y. Nakasuga and H. Ogawa (1995). Self-Heterodyning optical waveguide beam forming and steering network integrated on lithium niobate substrate, *IEEE Transactions on Microwave Theory and Techniques*, vol. 43, pp. 2395–2401.

[55] J. T. Gallo and R. DeSalvo (1997). Experimental demonstration of optical guided-wave butler matrices, *IEEE Transactions on Microwave Theory and Techniques*, vol. 45, pp. 1501–1507.

[56] J. Stulemeijer, F. E. van Vliet, K. W. Benoist, D. H. P. Maat and M. K. Smit (1999). Compact photonic integrated phase and amplitude controller for phased-array antennas, *IEEE Photonics Technical Letters*, vol. 11, pp. 122–124.

[57] M. A. Piqueras, S. De la Rosa, D. Zorrilla, J. Martí, G. Caille, J. M. Perdigués, P. Sanchis, A. Griol, B. Sanchez, G. Sanchez, J. V. Galán, A. Brimont, J. Hurtado and J. Ayucar (2007). Nanophotonic technology in space for beamforming applications, *6th ESA Round Table on Micro & Nano Technologies for Space Applications*, ESA/ESTEC, Noordwijk, The Netherlands, October 8–12.

[58] A. Meijerink *et al.* (2010). Novel ring resonator-based integrated photonic beamformer for broadband phased array receive antennas — Part I: Design and performance analysis, *Journal of Lightwave Technology*, vol. 28, pp. 3–18.

[59] D. Marpaung, L. Zhuang, M. Burla, C. Roeloffzen, J. Verpoorte, H. Schippers, A. Hulzinga, P. Jorna, W. P. Beeke3, A. Leinse, R. Heideman, B. Noharet, Q. Wang, B. Sanadgol and R. Baggen (2011). Towards a Broadband and Squint-Free Ku-Band Phased Array Antenna System for Airborne Satellite Communications, *5th European Conference on Antennas and Propagation*, Rome, Italy, April 11–15.

[60] M. Burla, C. G. H. Roeloffzen, L. Zhuang, D. Marpaung, M. R. Khan, P. Maat, K. Dijkstra, A. Leinse, M. Hoekman and R. Heideman (2012). System integration and radiation pattern measurements of a phased array antenna employing an integrated photonic beamformer for radio astronomy applications, *Applied Optics*, vol. 51, pp. 789–802.

[61] M. Burla, D. A. I. Marpaung, L. Zhuang, M. R. Khan, A. Leinse, W. Beeker, M. Hoekman, R. G. Heideman and C. G. H. Roeloffzen (2014). Multiwavelength-integrated optical beamformer based on wavelength division multiplexing for 2-D phased array antennas, *Journal Lightwave Technology*, vol. 32, pp. 3509–3520.

[62] R. L. Jordan (1980). The Seasat-A synthetic aperture radar system, *IEEE Journal of Oceanic Engineering*, vol. 5, pp. 154–164.

[63] J. C. Curlander and R. N. McDonough (1992). *Synthetic Aperture Radar: Systems and Signal Processing*. Wiley-Blackwell, Hoboken.

[64] C. D. Daniel (1986). Concepts and techniques for real-time optical synthetic aperture radar data processing, *IEEE Proceedings*, vol. 133, pp. 7–25.

[65] M. N. Armenise, E. Pansini and A. Fioretti (1988). A Novel Guided-Wave Correlator For Real-Time Synthetic Aperture Radar Data Processing, Proceedings SPIE, vol. 0993, 225.

[66] M. N. Armenise, F. Impagnatiello, V. M. N. Passaro and E. Pansini (1991). Design of a GaAs acousto-optic correlator for real-time processing, *Proceedings SPIE*, vol. 1562, 160.

[67] P. Bourqui, B. Harnisch, L. Marchese and A. Bergeron (2008). Optical SAR processor for space applications, *Proceedings SPIE*, vol. 6958, 69580J.

[68] L. Marchese, M. Doucet, B. Harnisch, M. Suess, P. Bourqui, N. Desnoyers, L. Guillot, F. Châteauneuf and A. Bergeron (2010). Introducing Real-Time On-Board SAR Image Generation using an Optronic SAR Processor, *8th European. Conference on Synthetic Aperture Radar (EUSAR)*, Aachen, Germany June 7–10.

[69] Y.-L. Desnos, C. Buck, J. Guijarro, J.-L. Suchail, R. Torres and E. Attema (2000). ASAR — Envisat's advanced synthetic aperture radar, *ESA Bullettin*, vol. 102, pp. 91–100.

Chapter 5

Image Detectors

The success of the space missions for Earth observation, monitoring of solar activity, Solar System exploration, and Universe investigation also depends on imaging detectors. The crucial role of these devices motivates the constantly growing research effort aimed at enhancing the performance of image sensors for space.

In this chapter, the technology of image detectors for many space applications is critically reviewed. Detectors operating in the range from ultraviolet (UV) to infrared (IR) are considered.

5.1 Visible image detectors: Charge-coupled device (CCD) sensors

Visible image detectors for space applications are silicon solid-state devices. When a photon with energy greater than the silicon energy-gap (1.12 eV at 300 K) impinges on the material, there is a certain probability that an electron–hole pair is generated [1]. Due to its electronic properties, silicon has photosensitive capability in the range of 400–1,100 nm.

An image detector consists of a bi-dimensional array of photosensitive cells (pixels). During a period of time known as exposure time or integration time, the array is exposed to light and accumulates electric charges inside its own pixels through the photo-transduction process. After photo-induced electrons are stored in each pixel, readout operations convert the charge signals into voltage signals, which hence can be successively processed by the electronics. The inverse of time between two following exposure operations determines the frame rate of the detector.

The major two classes of solid state image detectors in the visible spectrum are CCD and complementary metal–oxide-semiconductor (CMOS). Both of them are used today in space missions, although they are utilized in systems with different kinds of functionalities.

CCD detectors are preferred for applications requiring high resolution. Major examples are astronomy (e.g. Hubble Space Telescope [2] and GAIA mission [3]) and planetary remote sensing systems, in which the CCD detector is one of the fundamental blocks of spectrometers. The function of CCD detectors in spectrometers is to sense the visible light coming from a dispersive element, typically a diffraction grating, and after passing through different optical elements such as lenses and mirrors, going towards the surface of the CCD device itself.

On the other hand, using CMOS detectors all the advantages of CMOS technology are exploited. These rely on requirements about reduced power dissipation, mass, volume, and increased compactness and reliability of the overall imaging system, basically enabled by the integration capabilities offered by the technology itself. However, CMOS detectors generally lack photometric accuracy when compared with CCD devices due to their lower signal-to-noise ratio (SNR).

In this section, the operating principle of CCD detectors together with their performance parameters are presented. Typical operation modes of the imaging systems used in space applications are introduced. Then, after a brief introduction on the space environment, its impact on the performance parameters of CCD detectors is discussed.

5.1.1 *CCD fundamentals*

The basic building block of a CCD is the metal–oxide-semiconductor (MOS) capacitor formed by placing a conductive electrode (gate), insulated by a silicon dioxide film, over a p-doped silicon substrate [4]. The CCD consists of an 1D array of closely spaced MOS capacitors [5]. By applying appropriate voltages to the gates, the CCD is able to store and transfer the electrons that are generated by the photons absorption. The gate electrodes are connected together in groups of two, three or four

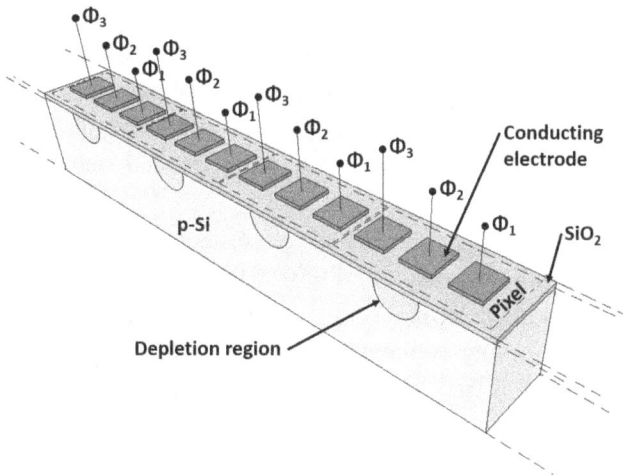

(a)

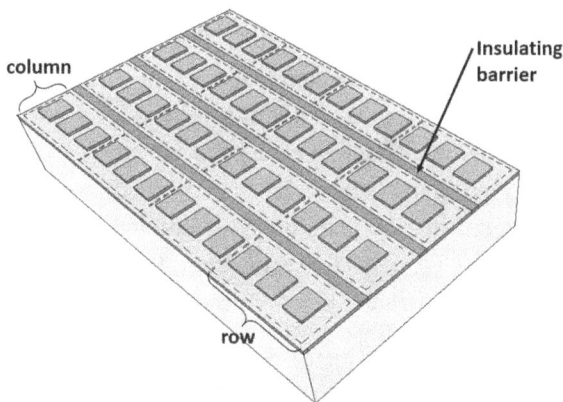

(b)

Fig. 5.1.   (a) Three-phase CCD. (b) 4 × 4 pixel array in a CCD image detector.

phases. The configuration of a three-phase CCD is shown in Fig. 5.1(a). Each pixel includes three closely spaced MOS capacitors whose gate electrodes are driven by the voltages Φ_1, Φ_2, and Φ_3. In a CCD image detector, thousands of CCDs, separated by insulating barriers, are located side by side, forming the columns of a 2D array of pixels (see Fig. 5.1(b)).

In a three-phase CCD image detector, during the exposure to photons, Φ_1, Φ_2, and Φ_3 are set so that there is a depletion region for each pixel, below one of the gate electrodes, for example the gate driven by the voltage Φ_2 (the typical values of Φ_1, Φ_2, and Φ_3 to achieve this condition are 5 V, 10 V, and 5 V, respectively). The photo-generated electrons are stored in the depletion region.

When the exposure time ends, an electromagnetically controlled shutter avoids other photons reaching the silicon substrate and the readout of the charge stored in each pixel starts. During the readout time Φ_1, Φ_2, and Φ_3 are varied according to three clock waveforms [5] so that the charge stored in each pixel is transferred to the adjacent pixel along each column. In this way, the charge stored in each row is transferred to the row below, as shown in Fig. 5.2. The charge in the lowest row is transferred in a serial CCD readout register, which is driven by a clock signal very faster than the three clock waveforms driving Φ_1, Φ_2, and Φ_3. In the readout register, at each clock pulse, the charge associated with each element of the register is transferred to the adjacent element. The charge associated with the register element directly connected to the output charge amplifier, is transferred to it and there it is transduced into a voltage level proportional to the charge at the amplifier input. Before each pixel can be readout, the node at the charge amplifier input must be reset by the so-called reset transistor. In this way, the charge stored in each pixel is transduced into a voltage level during the readout time. The CCD detector reconstructs the image by processing the voltage levels associated to each pixels in the 2D array.

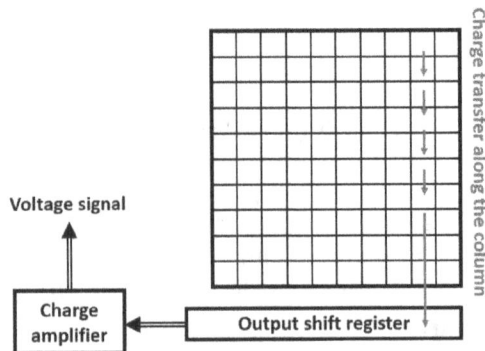

Fig. 5.2.   Configuration of a full-frame CCD image detector.

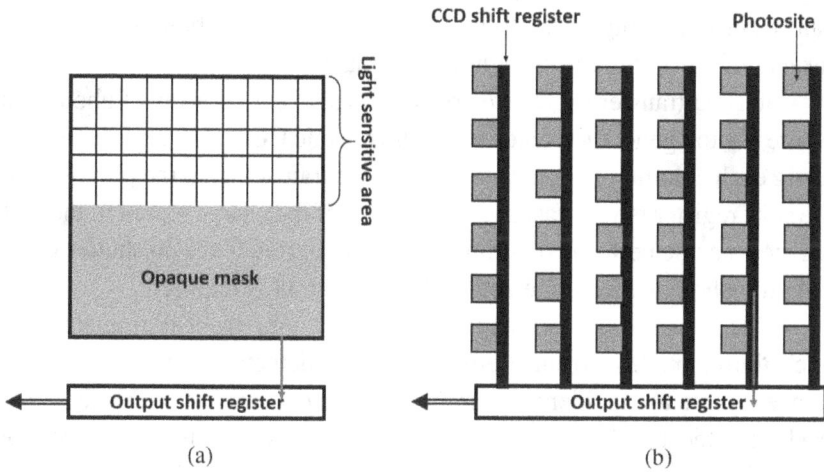

Fig. 5.3.   (a) Frame-transfer CCD image detector. (b) Interline-transfer CCD image detector.

The above-described CCD configuration, which is widely used in astronomy, is called full-frame CCD array [6]. The main drawback of this configuration is the quite low frame rate (typically of the order of 10 frames per second). To increase the frame rate, two alternative configurations have been developed: frame-transfer CCD (Fig. 5.3(a)) and interline-transfer CCD (Fig. 5.3(b)).

In frame-transfer CCD [6], half of the pixels are covered by an opaque mask avoiding the photons interacting with them, and the remaining pixels are exposed to the image and store photons during the exposure time. The former half of the capacitors array forms the frame store area while the latter one forms the image store area. This configuration allows a very quick shift of the charge stored during the exposure time from the image store area to the frame store area. After the end of the charge transfer from the image store area to the frame store area, the successive exposure time can start. The readout of the charge stored in the frame store area takes place during the exposure time, thus no shutter is required, so avoiding photon generation during the readout is required. The consequence of this operating mode is the increase in the frame rate with respect to the full-frame configuration. Since in frame-transfer CCD the exposure also continues during image dumping in the frame store area, image smears can occur (smearing refers to the appearance of a brighter vertical stripe instead of a bright point of light in the image).

Interline-transfer CCD detectors [6] consist of a 2D array of photosites, where photon generation takes place and opaque CCD shift registers placed along each column of photosites. The charge transfer from the photosites to the shift register is very fast and the readout of the charge stored in the shift registers can be performed during the exposure time. Thus no shutter inhibiting the photon/silicon interaction during readout is required.

CCD image detectors can be classified into front-illuminated and back-illuminated. In front-illuminated CCD detectors, photons from the imaged scene are incident on the front of the device, where the gate electrodes are located. In back-illuminated CCD detectors, light is incident on the back of the device. The fabrication process of front-illuminated detectors is easier and less expensive but the performance of these devices is limited by the interaction between the photons and the gate electrodes. By using transparent materials such as indium tin oxide for the gates, this critical aspect can be mitigated. Back-illuminated CCD detectors are usually preferred in those applications demanding high performance but their fabrication process includes an additional step, consisting of silicon substrate thinning, with a consequent increase in the cost of the device.

5.1.2 *Performance parameters*

The main performance parameters of CCD image detectors are readout noise, dynamic range, dark current, spectral sensitivity range, array size, and charge transfer efficiency (CTE).

Some noise sources can affect the analogue signal at the output of the CCD array, such as the thermal and the $1/f$ noise of the charge amplifier, and the kTC noise [7]. The kTC noise, also known as reset noise, consists of the thermal noise of the reset transistor, which is transferred to the charge amplifier during the reset operation. The name kTC comes from the expression of the root-mean-square (RMS) of the charge at the charge amplifier input due to this noise contribution. It is equal to $kTC^{1/2}$, with T being the absolute temperature, C the capacitance at the input node of the charge amplifier, and k the Boltzmann constant. To mitigate the kTC noise, a correlation technique, known as correlated double sampling

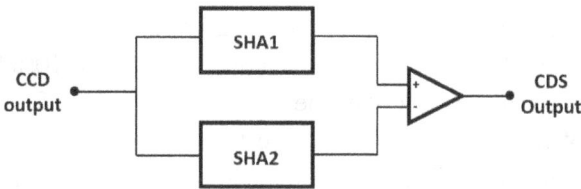

Fig. 5.4. Circuit implementation of the CDS technique.

(CDS) [8], is often utilized. An example of circuit implementation of the CDS technique is shown in Fig. 5.4. The CCD detector drives both the sample and hold amplifiers (SHAs), whose outputs are sent to a different amplifier. At the end of the reset interval, SHA1 holds the reset voltage level plus the kTC noise. At the end of the readout of each pixel, SHA2 holds the voltage level proportional to the charge stored in the pixel plus the kTC noise. In the difference between the outputs of the SHAs, the kTC noise is removed. After the CDS technique has been applied, the dominant noise contribution is the thermal noise of the readout amplifier, known as readout noise. The readout noise is white noise, so its variance increases linearly with the bandwidth of the amplifier, which is imposed by the frame rate of the detector. Today, high performing CCD devices show a readout noise of the order of a few electrons RMS per pixel [9].

Dynamic range is the range of charge, expressed in dB units, which an array can accumulate without distorting an image. The lower limit is represented by the noise floor superimposed on the signal, consisting of the readout noise, while the upper limit is represented by the largest amount of charge that can be stored in each pixel, known as the full well capacity (FWC). Currently, dynamic range values > 100 dB are achievable by high performance CCD detectors [9].

The dark current of CCD detectors refers to electron generation within the depletion region of each pixel not due to photo-generation. If the detectors are not cooled, the vast majority of the electrons are thermally generated. This is the reason why the CCD detectors for applications demanding low dark current are cooled to $-100°C$ or less. If the image sensor is properly cooled, the main sources of dark current are the surface traps at the Si/SiO_2 interface, which act as charge generation/recombination centers. Each pixel has its own contribution of dark current, so the global effect on the image appears as a statistical pixel-to-pixel non-uniformity, with RMS values typically ranging from 3% to 10% [10].

Spectral sensitivity describes the quantum efficiency of the detector as a function of the wavelength. Its range typically spans 400 nm to 1,100 nm. Front-illuminated CCD image detectors exhibit a low quantum efficiency (<10%) in the blue region of the electromagnetic spectrum because of the absorption of the polysilicon gates [11]. This critical aspect can be solved by using back-illuminated CCD image detectors, in which the silicon substrate is thinned, since they have a quantum efficiency >50% in the blue region [11].

CCD array size could be a critical parameter in some space applications, where imagers have to cover a large focal plane. The maximum size of a CCD pixel array is limited by the maximum reticle size of the photo-lithographic process in the manufacturing steps. Currently CCD image detectors with an image area of about 100 mm × 100 mm are available on the market [12]. To cover very large focal planes, it is possible to use several CCD detectors very close to each other (gap <0.5 mm).

In a CCD image detector, the CTE is defined as the fraction of electrons that are successfully moved from one pixel to the adjacent one during the charge transfer operation. In surface-channel CCD detectors, where the depletion regions are formed immediately beneath the thin silicon dioxide (as in Fig. 5.1(a)), interface traps at the Si/SiO_2 interface absorb and release charge with different time constants, resulting in the smearing of charge packets being transferred through the CCD. Consequently, the CTE is quite far from one in surface-channel CCD detectors. To improve the CTE, an additional n-type silicon layer can be implanted at the top of the p-type Si substrate, immediately below the silicon dioxide insulating layer. In this way, the depletion regions are formed deeper within the substrate and away from the traps at the Si/SiO_2 interface. The CTE of these buried-channel CCD detectors is typically very close to one but their FWC is lower with respect to surface-channel devices.

5.1.3 *Push-broom and time delay integration (TDI) imaging modes*

In some space applications, such as Earth and planetary remote sensing requiring the scanning of a scene, CCD imaging systems can be used in

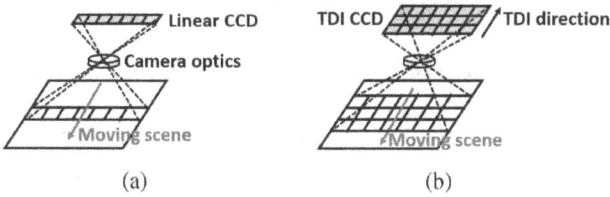

Fig. 5.5. (a) Linear push-broom imaging mode. (b) TDI imaging mode.

two operating modes: the linear push-broom imaging mode (see Fig. 5.5 (a)) and the TDI mode (see Fig. 5.5(b)).

In the linear push-broom imaging mode [13], a linear CCD is used with each element of the 1D array being interfaced with a photodiode through a transfer gate. The image is acquired by scanning the linear CCD across the target scene, transferring charges acquired by the photodiodes in the CCD elements, which are then serially transferred towards the charge-to-voltage conversion amplifier. The 2D image is formed by processing the 1D images acquired by the linear array during its motion. For example, the camera on the SPOT (*Satellite Pour l'Observation de la Terre*) satellites operates in this mode. It is based on a 6,000 pixel linear array covering an angle of 4.2°. Every 9 s it records 6,000 1D images, forming a 6,000 × 6,000 pixels image that covers a square region on the Earth's surface with a side of approximately 60 km [13].

The TDI operating mode [13] implies the use of imagers including a 2D array of CCD pixels and it is advantageous when the scene illumination is not sufficient to obtain a useful signal using an 1D array. The number of pixels in the array is large in the direction orthogonal to the scanning direction and relatively small in the other direction. During scanning, the charge stored in the pixels is transferred along the scanning direction, but in the opposite sense, towards the serial register connected to the output amplifier. The transfer rate is the same as the scanning rate. To avoid smearing, the charge transfer between the pixels and the movement of the scene are carefully synchronized. The TDI operating mode assures that the target scene is integrated over the number of lines in the array.

The High Resolution Imaging Science Experiment (HiRISE) camera on the Mars Reconnaissance Orbiter (MRO) captures images through 14 TDI CCD detectors, providing high resolution imaging [14].

5.1.4. *Radiation effects in CCD detectors*

The earth's magnetosphere is bombarded by a flux of energetic charged particles, formed by 85% protons (hydrogen nuclei), 14% α-particles (helium nuclei), and 1% heavier ions (the full range of elements, particularly carbon and iron nuclei) [15]. These charged particles are captured by Earth's magnetic field and have easier access to poles compared with the equator. This type of environment has great influence on the CCD performance parameters, with radiation effects generally grouped in: total ionizing dose (TID) effects and displacement damage effects, both degrading the CCD performance in a permanent way; and transient effects, which temporarily interfere with the nominal detector operation [16].

The TID effect on the MOS structure, on which CCD devices are based, is the generation of trapping site in the silicon dioxide insulating layer and at the Si/SiO_2 interface. This effect produces shifting of the flatband voltage that changes the effective biasing of the device and increases the surface recombination, with a consequent increase in dark current. In addition, the amplifier noise and its nonlinearity are also increased.

CDD performance in Space is not generally limited by TID effects, displacement damage effects being the most limiting factor. For TID above 5–10 krad (Si), changes in performance of the output amplifier and shifts in the clock voltage are observed [17]. However, the device functionality holds up to several tens of krad (Si). More radiation resistant behavior can be achieved by thinning the dielectric layer and by balancing the electron and hole trapping using dual oxide/nitride dielectrics, holding the device functionality up to 1 Mrad (Si) [17].

Displacement damage effects are caused by collision of heavy particles, such as protons and neutrons, with silicon atoms, which are ultimately displaced from their lattice sites and occupy interstitial sites, that behave as defects. Some of them recombine with each other, but the remaining ones form stable defects such as the phosphorus-vacancy complex (called E-centers) and oxygen-vacancy defects (called A-centers). These defects degrade the CCD performance parameters by decreasing CTE, increasing the mean value and spatial non-uniformity of dark current, generating individual pixels with very high dark currents (hot pixels) [17].

As shown by the XMM-Newton and Hubble Space Telescope (HST) programs, proton-induced CTE degradation could be a critical parameter for astronomy tasks involving acquisition of very faint objects on a dark background [18]. As the array format increases, the number of transfer operations through the detector columns increases. Moreover, the number of transfers experimented by a charge packet depends on its position in the array. Upper positions, far from the serial readout register, are subject to many more transfers. These packets lose more electrons than those located near the register and could be overwhelmed by the charge amplifier read-out noise. This means that the photometric accuracy varies across the proton-irradiated detector, being higher for stars and galaxies imaged near the serial register. In the Advanced Camera for Surveys on board the HST, this critical aspect can be prevented by flashing the CCD array after charge integration and before charge readout with an optical source allowing the filling of radiation-induced traps. This technique is only recommended in certain specific scientific observations, e.g. very faint and unresolved/compact targets [19].

In summary, displacement damage effects often dominate the radiation response in the state-of-the-art scientific detectors when they operate in the space environment. A significant CTE degradation has been observed for proton exposure less than 1 krad (Si).

Transient effects occur when particles such as protons or cosmic rays pass through the active volume of a CCD device, ionization-induced charges appear along the entire path of impinging particles, producing tracks that may cross many pixels [20]. These induced charges are then transferred during CCD operation to the readout register, causing interference in the image. Interferences can be treated as noise, which can be reduced by using a Kalman filter.

5.1.5 *Examples of space missions employing CCD detectors*

As already mentioned, CCD image detectors are widely used in astronomy observatories where faint sources are observed, generally on a dim background and characterized by a low flux of photons. In this application domain, CCD detectors with medium/long integration times and very low noise are required. A recent example of the CCD imager application is a space-based astronomy mission by the GAIA mission, which aims at drawing a 3D map of our Galaxy. The mission is based on a large array of 106 close-butted CCD detectors [21], operating at $-110°C$, each of them including $1,966 \times 4,500$ pixels with size $30 \times 10 \ \mu m^2$.

CCD image detectors are also utilized in Earth or planetary observation missions, when spatial and spectral variations of brighter scenes involving high flux of photons should be monitored. In this case, short integration times, large FWC, high dynamic range, and low noise are required for the CCD imager. An example of a high-flux space application of CCD image sensors is given by the spectrometer developed for the Sentinel-4 mission, which will monitor the composition of the Earth's atmosphere above Europe. The CCD detectors selected for that application are thinned frame-transfer CCD devices [22].

5.2 Visible image detectors: CMOS sensors

Along with CCD sensors, CMOS detectors [23] are used in space applications for imaging in the visible range. CMOS technology is used in the fabrication of detectors formed by an array of photosensitive elements, readout and control electronics, and eventually A/D conversion electronics all integrated on the same chip, thus decreasing the overall imaging system mass and volume, and increasing its compactness and reliability. Furthermore, while charge transfer in CCD devices requires high and highly synchronized control voltages, of about 10 V, CMOS image sensors require operating voltages of a few volts, with a consequent reduction in the power consumption down to tens of mW.

In CMOS detectors, image sensing is based on the photo-transduction process that takes place in silicon when interacting with photons having a wavelength in the visible range. In such image sensors, the opto-electronic device that produces the electrical signal proportional to the amount of light impinging on it is a reverse biased $p–i–n$ silicon photodiode. The detector consists of a bi-dimensional array of pixels each containing a photodiode, acting as a photosensitive element, with the relative readout electronics (see Fig. 5.6). The image detector operates similarly to a random access memory, wherein all the pixels in the same row are addressed

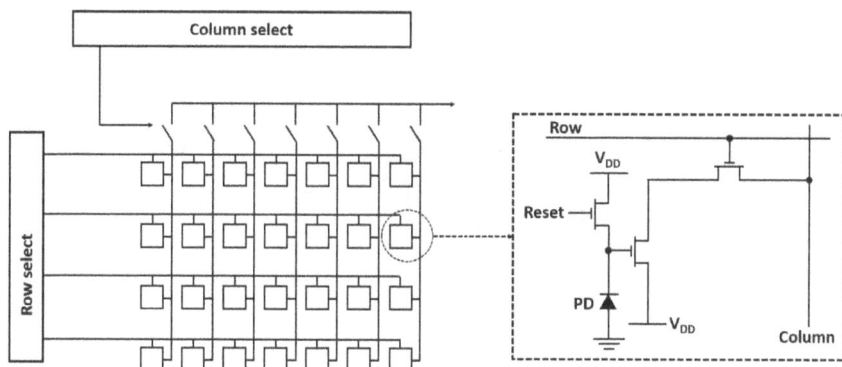

Fig. 5.6. CMOS image detector. The inset shows the 3T pixel configuration.

simultaneously by the row select lines. When a row of pixels is active, voltages at the pixel output are put on the column lines and sampled. By selecting all the rows in turn, all the pixels can be sampled and the image can be reconstructed.

A quite typical pixel configuration in a CMOS image sensor is the 3T one [24] (see the inset in Fig. 5.6), including a photodiode and three transistors for readout operations.

Before the detector exposure, the M_{RS} transistor sets the photodiode voltage to a pre-fixed value, the reset value, and then leaves the photodiode in a floating condition. During the exposure time, visible photons impinging on the photodiode generate electron–hole pairs in the silicon, thus discharging the capacitance associated with the photodiode itself. When the row select line activates the M_{SEL} transistor, the source follower (M_{SF}) operates a charge-to-voltage conversion and the signal is transferred to the column output line.

Since in CMOS detectors the first signal conditioning is performed within each pixel, these detectors are also called active pixel sensors (APSs).

In CMOS detectors, each pixel has its own readout electronics with its relevant noise contributions, including $1/f$ noise and thermal noise of transistors, and kTC noise of the reset transistor. The latter noise contribution can be mitigated by using an alternative pixel configuration including four transistors and thus called 4T [24] (see Fig. 5.7). In this configuration, the

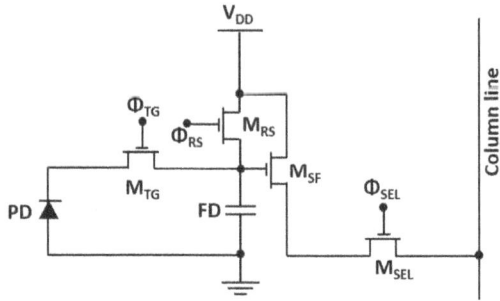

Fig. 5.7.   4T pixel configuration in CMOS image detectors.

accumulated photo-generated carriers are transferred to a capacitive floating diffusion (FD), where the carriers are converted to a voltage. This transfer is controlled by the M_{TG} transistor. The FD is reset after the carriers' photo-generation in the PD and just before the pixel readout. After the FG reset, the signal charge accumulated in the PD is transferred to the FD and then readout by turning on M_{SEL}. In this way the FD and not the PD is reset, with a consequent reduction in the kTC noise.

Imagers based on both the 3T and 4T configurations are affected by spatial noise, due to the mismatch between pixels, which are not at all identical.

All the noise sources contribute to the detector SNR, which is generally worse in CMOS detectors than in CCD ones. Although a large amount of noise can also be suppressed in CMOS devices using the CDS technique, the photometric accuracy of CMOS detectors cannot compete with that of CCDs, which remain the preferred image sensors for high resolution space applications due to their high immunity to noise.

However, the above-mentioned advantages of CMOS technology are very attractive for space applications where reduced mass, volume, and power consumption are key parameters. Furthermore, the system integration capability offered by CMOS technology enables the realization of very compact systems. All this makes CMOS detectors very appealing for some sub-systems, such as star trackers, which are used for spacecraft attitude control [25]. The possibility of utilizing CMOS detectors in terminals for optical satellite/aircraft communication links has also been evaluated [26].

5.2.1 *Performance parameters and radiation effects in CMOS detectors*

Performance parameters of CMOS image detectors are almost the same as those already introduced for CCD detectors. The noise source are also very similar and the CDS technique is again used to reduce the noise introduced by the pixel electronics. In CMOS detectors, the CDS circuitry is integrated on-chip as part of the readout electronics, and placed at the end of each column-line of the pixel matrix.

The dynamic range depends, as in the CCD case, on the largest charge that can be stored inside the pixel, and on the readout noise at the output stage. Although different enhancement techniques have been implemented, CMOS detectors generally show a lower dynamic range when compared with their CCD counterparts.

The spectral efficiency is generally affected by the low fill factor (FF) of CMOS detectors. The FF is defined as the ratio between the photosensitive area of the pixel and its total area, and indicates the fraction of photons effectively converted by the pixel itself. The photons that do not impinge on the photosensitive area are absorbed by the in-pixel transistors or reflected by the aluminum interconnection tracks. The solution to overcome this issue has been borrowed from CCD devices, and consists of thinning and back-illuminating the device to improve its quantum efficiency over the whole spectral range. This solution also has the capability of extending spectral sensitivity range of the device [27].

The most attractive feature of CMOS detectors for space applications is the low power consumption. In fact, CMOS devices have a power consumption much lower than that of CCD ones, especially if the conversion and processing electronics is taken into account in the evaluation.

As with CCD devices, CMOS image detectors also suffer from TID and displacement damage effects. In addition, CMOS detectors are susceptible to a potentially destructive condition referred to as Single-Event Latch-up (SEL) [28]: a single particle such as a proton, neutron, or heavy ion, when interacting with the silicon bulk can trigger, thanks to a parasitic thyristor formed intrinsically in CMOS structures, an over-current phenomenon that could lead to the destruction of the device. Nominal operating conditions can only be recovered by temporarily removing the power supply.

CMOS image detectors generally show better behavior with respect to CCD ones in radiation environments for both TID and displacement damage effects. This feature is obtained by selecting specific design techniques in the standard CMOS process that makes the detector more radiation-tolerant, which is a great advantage for missions close to the Sun such as BepiColombo and Solar Orbiter [29].

5.2.2 *Examples of CMOS-based systems in space missions*

As already mentioned, CMOS detectors cannot compete with CCD ones in terms of photometric accuracy, low noise, and high dynamic range capabilities, although they posses very attractive features in terms of low power, size, and weight. An example of a space system that notably benefits from the features of CMOS imagers is the star tracker, that is widely used in spacecraft attitude control.

A star tracker [30] is a fully autonomous attitude sensor, acquiring digital images of the sky, identifying the stars visible in that image and calculating the attitude and the angular rate of the spacecraft with high accuracy. It operates by acquiring a full format image of the portion of sky contained in its field of view and processing it to observe stars, whose centroids are calculated. The star map is then matched with those of an on-board database and, in this way, the spacecraft attitude is computed.

When the star tracker is in its tracking mode, searching and evaluating the position of stars, it operates computations on different regions of interest within the image, e.g. for centroid calculation. The random access feature of CMOS imagers enables the data transfer of selected regions of interest to the downstream processor, providing a great reduction in bandwidth requirements of the readout electronics and thus reducing noise and costs. Furthermore, all the data conditioning and conversion circuitry is integrated on the same sensor chip, reducing the mass, volume, weight and power consumption of the overall star tracker.

A star tracker with noise equivalent angle (RMS noise observed in successive attitude updates when the instrument is acquiring the same image) less than 0.8 arcsec/$\sqrt{Hz}$ was launched on board the French Spot-6 Earth observation satellite on September 9, 2012. During its operation, the sensor had the same performance that was estimated by the ground tests [31].

During recent years, the European Space Agency (ESA) has been developing, with some industrial partners, CMOS imagers aimed at replacing CCD ones in high-resolution applications. Sentinel-2 is an example of an Earth observation mission in which CMOS image detectors will have a key role. In the framework of that mission, two identical satellites will be launched in April 2015 and 2016. Each satellite contains the Multi Spectral Instrument (MSI), acquiring images in 13 spectral bands ranging from the visible and the near IR to the short wave infrared. The image detectors operating in the visible and the near infrared bands are based on CMOS technology [32].

5.3 UV image detectors

UV image detectors are typically used in space instrumentation for solar UV astronomy and weather space monitoring, which rely on the monitoring of solar activity and its effect on the Earth's environment. Solar UV astronomy studies the solar activity through its irradiance spectrum analysis, with particular interest in the investigation of solar phenomena such as solar flares and corona mass ejection [33]. Weather space monitoring concerns more general phenomena outside the Earth's atmosphere and their effects on the terrestrial environment. For example, the Solar Dynamics Observatory (SDO) satellite [34], which was launched by NASA on February 11, 2010 to monitor the topology of solar magnetic fields, includes an array of four telescopes called the Atmospheric Imaging Assembly (AIA) that provides continuous observations of the solar chromosphere and corona at several UV wavelengths.

5.3.1 *UV image detectors classification*

The atmosphere surrounding the Earth prevents UV radiation from the Sun having wavelength <290 nm reaching the Earth's surface. This behavior is due to absorption of different molecules, especially oxygen and ozone, present in the Earth's atmosphere. UV detectors not sensitive to UV radiation with wavelength >290 nm are referred to as solar-blind, because they are only responsive to UV wavelengths shorter than that of solar radiation penetrating the Earth's atmosphere. UV detectors that are only sensitive to wavelengths <400 nm, the lower edge of human eye sensitivity, are called visible-blind.

Apart from this first wavelength-based classification, UV image detectors can be classified in two main classes: photo-emissive detectors and photo-conductive detectors [35].

In photo-emissive detectors, a cathode coated with a material emitting electrons when UV radiation hits the surface is utilized. After the ejection, the single photo-induced electron needs some form of amplification to be recorded in a reliable way. This amplification is obtained by an image intensifier that generates a cascade of electrons *via* secondary emission. The secondary electrons at the intensifier output are readout by an electronic detector. It records the positions of electrons impinging on it, allowing the image to be reconstructed. The choice of the material for the cathode coating defines the operating wavelength range of the detector. Photo-emissive imaging systems are typically characterized by very low noise and high gain, which enables a good photon-counting capability.

The basic feature of photo-conductive detectors is the broad operating wavelength range. Photo-conductive sensors absorb photons into a semiconductor region where electron transitions from the valence to the conduction band occur. The most common UV photo-conductive detectors are the silicon-based CCD devices, which are typically visible-blind detectors. In present years, a notably R&D effort has been dedicated to reducing the readout noise and enhancing the UV quantum efficiency of CCD detectors. Currently CCD devices have an SNR that is competitive with that of photo-emissive detectors. Furthermore, CCD detectors are compact and more reliable with respect to their photo-emissive counterparts. The main critical aspects of most of the CCD sensors available on the market are the high quantum efficiency in visible wavelengths and the low quantum efficiency in UV.

Silicon, due to its band gap, is not the ideal material for UV image detectors. This is the reason why alternative semiconductors, the wide band gap (WBG) ones, are currently under investigation for the development of a new generation of UV detectors [35]. In WBG detectors, visible-blindness is observed when the semiconductor has a band gap >3.4 eV, such as GaN, while solar-blindness is achieved with an energy band gap of at least 4.2 eV, which can be obtained by varying the Al fraction in the AlGaN alloy. Some prototypes of solar-blind image detectors based on AlGaN *p–i–n* photodiode arrays, with peak quantum efficiency in the UV wavelength range up to 66%, have been demonstrated in last decade [36, 37]. In this prototype, the readout electronics is implemented using CMOS technology.

5.3.2 *MCP-based and CCD detectors*

The most common photo-emissive UV detector in space applications is based on the microchannel plate (MCP) structure [38]. The MCP consists an array of microchannels, roughly 10 μm in diameter, fabricated in 1 mm thick glass plate. Both faces of the plate are coated with thin metal films acting as electrodes across which a voltage is applied. The inner wall of each channel is coated with an emitting material so that, when a photo-induced electron enters the channel, electron multiplication takes place and a cloud of 10^4 to 10^6 electrons exits from the plate. Image can be reconstructed thanks to an electronic grid, which estimates the number of electrons imprinting in each point of the 2D grid. MPC-based detectors need a pressure $<10^{-6}$ torr to properly operate. Figure 5.8 shows a schematic cross-section of the MCP structure.

MCP-based UV detectors allow very low noise and high gain operations, enabling photon-counting capability. Solar-blind detectors can be fabricated by using an appropriate photo-cathode material. Currently, MCP detectors have a UV quantum efficiency of about 10%. Requirements on the electron intensification imply high voltage operations, with a consequent decrease in system reliability, as well as an increase in size and weight. Additionally, unlike CCD detectors, MCP devices exhibit a quantum efficiency that is the function of the photon angle of incidence.

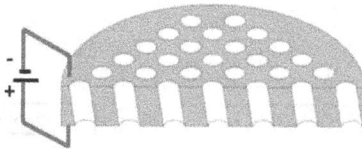

Fig. 5.8.   Schematic cross-section of MCP structure.

Imaging systems based on MCP structure are used in many space instruments operating in the UV wavelength range [39].

To be competitive with MCP-based detectors, CCD UV detectors should exhibit high and stable quantum efficiency at UV wavelengths and low noise. The thinned back-illuminated configuration is the preferred one in CCD detectors operating in the UV range. A back-illuminated CCD detector, in which the back surface is properly treated by using molecular beam epitaxy, having an internal quantum efficiency close to 100% in the visible and UV, was developed at the Jet Propulsion Laboratory at the beginning of the 1990s [40].

The performance of UV CCD detectors can be improved by modifying the conventional detector configuration, adding a multiplication register at the output of the readout CCD shift register. This detector configuration is called Electron Multiplying CCD (EMCCD) [41]. Figure 5.9 shows an electron multiplying version of a frame transfer CCD. The multiplication register is driven by relatively high voltages (up to 50 V). In this way, electron multiplication induced by impact ionization takes place in each stage of the multiplication register. Each stage amplifies the charge packets by a factor of about 1.01. Thus, for example, a multiplication register with 600 stages provides an overall signal gain of roughly 390 ($=1.01^{600}$). An EMCCD detector with an external quantum efficiency $>50\%$ in the UV range from 100 nm to 300 nm has been recently demonstrated [42]. The back surface of that detector, where the

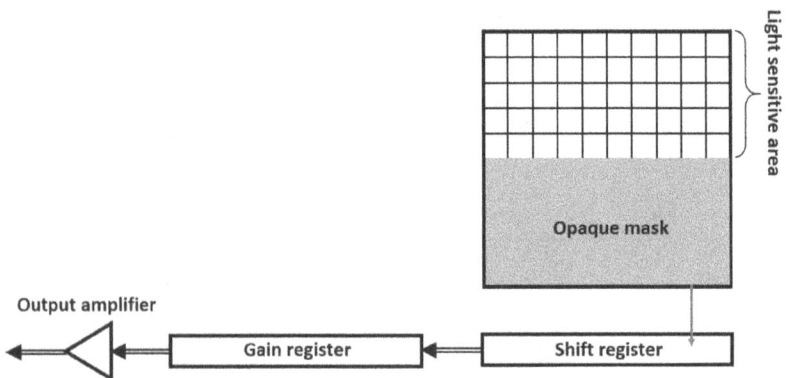

Fig. 5.9.   Electron multiplying CCD.

photons impinge, is properly formed by molecular beam epitaxy and atomic layer deposition.

5.4 IR image detectors

IR imaging provides great information about a planet's atmosphere since many chemical species have spectral signatures in IR bands, due to absorption processes associated with molecule vibrational states. IR detectors for space are solid-state devices quite similar to the previously discussed visible image detectors and solid-state UV image detectors. The most significant difference relies on the active material used for the detector fabrication.

The IR band covers a wide wavelength interval between visible and microwave bands, it is subdivided into many different IR sub-bands, i.e. NIR (near-IR, from 0.75 μm to 1.4 μm), SWIR (short-wavelength IR, from 1.4 μm to 3 μm), MWIR (mid-wavelength IR, from 3 μm to 8 μm), LWIR (long-wavelength IR, from 8 μm to 15 μm), and FIR (far IR, from 15 μm to 1,000 μm).

IR image detectors typically have a hybrid structure consisting of a photodiode array (focal plane array, FPA) made of a material absorbing photons at the IR wavelengths, and a silicon CMOS readout circuit, called readout integrated circuit (ROIC) [43]. The silicon ROIC implements the same signal processing techniques already introduced for visible detectors, including the use of the CDS technique for noise.

In space applications, the most used active material for IR image detectors is the MCT or HgCdTe [44], a ternary semiconductor compound containing Mercury, Cadmium, and Tellurium (MCT). By tuning the Cadmium fraction, the absorption peak of the MCT can be varied from near-infrared (NIR) to very long wavelength infrared (VLWIR).

The band gap of the MCT is direct, with a consequent high absorption coefficient and quantum efficiency of the detectors. In addition, molecular beam epitaxy and metal-organic chemical vapor deposition make it possible to produce large-diameter wafers having a top epitaxial layer of MCT with low density of defects and reproducible stoichiometric properties. The choice of the best substrate material on which the epitaxial layer of MCT can grow is an open field in the research community, with the aim of identifying materials allowing low cost, low defect density and large detector size [45]. The most used substrate material is CdZnTe (Cadmium Zinc Telluride, also known as CZT), a crystalline material with lattice structure similar to that of MCT. Currently, the main alternatives to CZT are GaAs, Si, and Ge. Silicon has several advantages with respect to CZT, such as larger array area, scalability, greatly reduced cost per unit area, good compatibility with the silicon ROIC, better robustness, existence of many substrate suppliers, higher quality (lower defect densities and less impurities). The main disadvantage is the lattice mismatch between silicon and MCT, which is a significant challenge for large FPA production.

The ROIC performance has great influence on that of the IR image detector since the small amounts of charge generated by MCT pixels are immediately transferred to ROIC electronics. Each MCT pixel faces a ROIC cell containing the single pixel conditioning electronics, as in CMOS detectors. For signal transfer, the MCT array and the ROIC are hybridized with the bump bonding [46].

The main challenge with hybrid detectors relies on the different material properties of the two components, the MCT array and the ROIC, so that cooling the detector to achieve a low operating temperature can cause the device to break due to the different thermal behavior of the materials. In fact, for space applications, IR detectors are typically cooled to approximately 80 K in order to reduce background noise, including the dark current.

Even though MCT technology is the most used for IR detectors, alternative technologies in different IR bands are under evaluation by the research community with the goal of achieving low dark current, low detector cost, thermal stability, radiation hardness, fast response, and low power consumption. For example, in the SWIR band, InGaAs detectors [47] are widely used, also in space missions [48], while in the MWIR and the LWIR bands quantum well infrared photo-detectors [49, 50] are under investigation. In addition, microbolometers in silicon MEMS technology [51], which are thermal IR detectors, have become commercial devices. They consist of an array of resistive elements at micro-scale that absorb electromagnetic radiation and thus increase their temperature. The microbolometers can be easily integrated with the ROIC.

References

[1] D. Durini (2014). Operating Principles of Silicon Image Sensors, in *High Performance Silicon Imaging: Fundamentals and Applications of CMOS and CCD Image Sensors*, D. Durini (Ed.), Woodhead Publishing, Cambridge.

[2] E. C. Sutton (2012). *Observational Astronomy: Techniques and Instrumentation*. Cambridge University Press, Cambridge.

[3] C. R. Kitchin (2014). *Astrophysical Techniques*, *6th* Edition, CRC Press, Florida.

[4] S. M. Sze and K. K. Ng (2007). *Physics of Semiconductor Devices*, 3rd Edition. John Wiley & Sons, New Jersey.

[5] M. J. Howes and D. V. Morga (Eds.) (1979). *Charge-coupled Devices and Systems*. John Wiley & Sons, New Jersey.

[6] S. S. Ipson and C. Okereke (2014). *Vision and Image Sensors*. CRC Press, Florida.

[7] A. J. Theuwissen (2002). *Solid-State Imaging with Charge-Coupled Devices*. Kluwer Academic Publishers, Dordrecht, The Netherlands.

[8] W. Kester (2005). Sensor Signal Conditioning, in *Sensor Technology Handbook*, J. S. Wilson (Ed.), Newnes.

[9] P. Garé, N. Nelms, Y. Nowicki-Bringuier, D. Martin, R. Meynart and M. Zahir (2013). Requirements, developments and challenges for CCD and CMOS image sensors for space applications, *2013 International Image Sensor Workshop*, Snowbird Resort, Utah, USA, June 12–16.

[10] N. Waltham (2013). CCD and CMOS Sensors, in *Observing Photons in Space: A Guide to Experimental Space Astronomy* M. C. E. Huber, A. Pauluhn, J. L. Culhane, J. G. Timothy, K. Wilhelm, A. Zehnder (Eds.), Springer, New York.

[11] M. Suyama (2009). Optoelectronic Sensors, in *Handbook of Optical Metrology: Principles and Applications*, T. Yoshizawa (Ed.), CRC Press, Florida.

[12] Datasheet of the CCD290-99, http://www.e2v.com/.

[13] G. Joseph (2015). *Building Earth Observation Cameras*. CRC Press, Florida.

[14] A. S. McEwen *et al.* (2007). Mars reconnaissance orbiter's high resolution imaging science experiment (HiRISE), *Journal of Geophysical Research*, vol. 112

[15] L. Miroshnichenko (2012). *Radiation Hazard in Space*. Kluwer Academic Publishers, Dordrecht, The Netherlands.

[16] C. J. Marshall (2012). Photonic Devices, in *Extreme Environment Electronics*, J. D. Cressler and H. A. Mantooth (Eds.), CRC Press, Florida.

[17] C. J. Marshall and P W. Marshall (2003). CCD Radiation Effects and Test Issues for Satellite Designers, NASA-GSFC Multi-Engineering Disciplinary Support Contract Task 1058.

[18] J. C. Pickel, A. H. Kalma, G. R. Hopkinson and C. J. Marshall (2003). Radiation effects on photonic imagers — A historical perspective, *IEEE Transactions on Nuclear Science*, vol. 50, pp. 671–688.

[19] S. Ogaz, M. Chiaberge and N. A. Grogin (2014). Post-Flash Capabilities of the Advanced Camera for Surveys Wide Field Channel (ACS/WFC), Instrument Science Report ACS 2014-01.

[20] G. R. Hopkinson and A. Mohammadzadeh (2004). Radiation effects in charge coupled device (CCD) imagers and active pixel sensors, *International Journal of High Speed Electronics and Systems*, vol. 14, pp. 135–443.

[21] J. H. J. de Bruijne (2012). Science performance of Gaia, ESA's space-astrometry mission, *Astrophysics and Space Science*, vol. 341, pp. 31–41.

[22] G. B. Courrèges-Lacoste, B. Ahlers, B. Guldimann, A. Short, B. Veihelmann and H. Stark (2011). The Sentinel-4/UVN instrument on-board MTG-S, EUMETSAT Meteorological Satellite Conference, Oslo, Norway, September 5–9.

[23] O. Yadid-Pecht and R. Etienne-Cummings (Eds.) (2004). *CMOS Imagers: From Phototransduction to Image Processing.* Kluwer Academic Publishers, Dordrecht, The Netherlands.

[24] A. El Gamal and H. Eltoukhy (2005). CMOS image sensors, *IEEE Circuits and Devices Magazine*, vol. 21, pp. 6–20.

[25] J. Bogaerts (2014). Complementary metal-oxide semiconductor (CMOS) image sensors for use in space, in *High Performance Silicon Imaging: Fundamentals and Applications of CMOS and CCD Image Sensors*, D. Durini (Ed.), Woodhead Publishing, Cambridge.

[26] O. Saint-Pé, M. Tuleta, R. Davancensa, F. Larnaudiea, P. Magnanb, P. Martin-Gonthierb, F. Corbièreb and M. Estribeaub (2005). Space optical instruments optimization thanks to CMOS image sensor technology, *Proceedings of SPIE*, vol. 5978, p. 597811.

[27] N. R. Waltham, M. Prydderch, H. Mapson-Menard, P. Pool and A. Harris (2007). Development of a thinned back-illuminated CMOS active pixel sensor for extreme ultraviolet spectroscopy and imaging in space science, *Nuclear Instruments and Methods in Physics Research A*, vol. 573, pp. 250–252.

[28] V. Lalucaa, V. Goiffon, P. Magnan, G. Rolland and S. Petit (2013). Single-event effects in CMOS image sensors, *IEEE Transactions on Nuclear Science*, vol. 60, pp. 2494–2502.

[29] P. Martin-Gonthier, P. Magnan and F. Corbiere (2005). Overview of CMOS process and design options for image sensor dedicated to space applications, *Proceedings of SPIE*, vol. 5978, p. 597812.

[30] W. Ley and F. Merkle (2009). Subsystems of spacecraft, in *Handbook of Space Technology*, W. Ley, K. Wittmann and W. Hallmann (Eds.), John Wiley & Sons, New Jersey.

[31] CMOS Star Trackers: Hydra Family. Available at: http://www.sodern.com/sites/en/ref/Star-Trackers-HYDRA_51.html.

[32] P. Martimort, V. Fernandez, V. Kirschner, C. Isola and A. Meygret (2012). Sentinel-2 MultiSpectral imager (MSI) and calibration/validation, *IEEE International Geoscience and Remote Sensing Symposium (IGARSS)*, Munich, Germany, July 22–27.

[33] A. Bhatnagar, W. Livingston, W. C. Livingston. *Fundamentals of Solar Astronomy*. John Wiley & Sons, New Jersey.

[34] P. Chamberlin, W. D. Pesnell and B. Thompson (2012). *The Solar Dynamics Observatory*. Springer, New York.

[35] P. E. Malinowski (2011). *III-N Ultraviolet Detectors for Space Applications*, PhD Thesis, University of Leuven.

[36] S. Aslam, F. Yan, D. E. Pugel, D. Franz and L. Miko (2005). Development of ultra-high sensitivity wide-band gap UV-EUV detectors at NASA Goddard Space Flight Center, *Proceedings of SPIE*, vol. 5901, pp. 59011J.

[37] E. Cicek, R. McClintock, A. Haddadi, W. A. G. Rojas and M. Razeghi (2014). High Performance Solar-Blind Ultraviolet 320×256 Focal Plane Arrays Based on $Al_xGa_{1-x}N$, *IEEE Journal of Quantum Electronics*, vol. 50, pp. 593–597.

[38] J. Vallerga, O. Siegmund, A. Tremsin and J. McPhate (2011). Current and future capabilities of MCP detectors for UV-VIS instruments, Cosmic Origins Program Analysis Group (COPAG) Workshop, Baltimore, MD, USA, September 22–23.

[39] O. H. W. Siegmund (2011). MCP Based UV Detectors, Their Evolution Through Many Astrophysics Missions and Their Future Scientific Applications, *AAS 217th Meeting*, Seattle, Washington, USA, January 9–13.

[40] M. E. Hoenk, P. J. Grunthaner, F. J. Grunthaner, R. W. Terhune, M. Fattahi and H.-F. Tseng (1992). Growth of a delta-doped silicon layer by molecular beam epitaxy on a charge-coupled device for reflection-limited ultraviolet quantum efficiency, *Applied Physics Letters*, vol. 61, p. 1084.

[41] D. J. Denvir and E. Conroy (2003). Electron multiplying CCDs, *Proceedings of SPIE*, vol. 4877, p. 55.

[42] S. Nikzad, M. E. Hoenk, F. Greer, B. Jacquot, S. Monacos, T. J. Jones, J. Blacksberg, E. Hamden, D. Schiminovich, C. Martin and P. Morrissey (2012). Delta-doped electron-multiplied CCD with absolute quantum efficiency over 50% in the near to far ultraviolet range for single photon counting applications, *Applied Optics*, vol. 51, pp. 365–369.

[43] I. M. Backer (2011). HgCdTe photovoltaic Infrared detectors, in *Mercury Cadmium Telluride: Growth, Properties and Applications*, P. Capper and J. Garland (Eds.), John Wiley & Sons, New Jersey.

[44] M. A. Kinch (2007). *Fundamentals of Infrared Detector Materials*. SPIE.

[45] J. Garland and R. Sporken (2011). Substrates for the Epitaxial Growth of MCT, in *Mercury Cadmium Telluride: Growth, Properties and Applications*, P. Capper and J. Garland (Eds.), John Wiley & Sons, New Jersey.

[46] P. Garrou, J. Jian-Quiang and P. Ramm (2012). Three dimensional integration, in *Handbook of Wafer Bonding*, P. Ramm, J. Jian-Qiang Lu and M. M. V. Taklo, John Wiley & Sons, New Jersey.

[47] A. K. Sood, Y. R. Puri, N. K. Dhar and D. L. Polla (2015). Recent advances in EO/IR imaging detector and sensor applications, in *Handbook of Sensor Networking: Advanced Technologies and Applications*, J. R. Vacca (Ed.), CRC Press, Florida.

[48] J. Bentell *et al.* (2010). 3000 pixel linear InGaAs sensor for the Proba-V satellite, *Proceedings of SPIE*, vol. 7862, p. 78620.

[49] K. K. Choi (1997). *The Physics of Quantum Well Infrared Photodetectors*. World Scientific Publishing, Singapore.

[50] A. Berurier, A. Nedelcu, V. Guériaux, T. Bria, A. de Rossi, X. Marcadet and P. Bois (2011). Optimization of broadband (11–15 μm) optical coupling in quantum well infrared photodetectors for space applications, *Infrared Physics & Technology*, vol. 54, pp. 182–188.

[51] F. Niklaus, C. Vieider and H. Jakobsen (2007). MEMS-based uncooled infrared bolometer arrays — A review, *Proceedings of SPIE*, vol. 6836, p. 68360D.

Chapter 6

Photonic Sensors and Instruments

Spacecrafts include a wide number of sensors monitoring several physical parameters and functionalities of their subsystems. For example, in a telecom satellite, hundreds of temperature sensors are typically installed and the attitude estimation requires a very sophisticated sensing system. Several enabling technologies, including photonics, are currently utilized to fabricate sensing devices/sub-systems, which have to be compliant with the stringent requirements imposed by the space application. Typical features of photonic sensors, e.g. good resistance to radiation and high temperature gradients, immunity to electromagnetic interference, and compactness, match very well with those requirements. Therefore, fiber and, more recently, microphotonic sensors are finding more and more space applications from angular velocity sensing, where they have been routinely used for several years, to new application domains such as temperature or strain mapping.

Light is a powerful tool for *in situ* and remote sensing. Consequently, photonic instruments are very useful payloads for Earth observation and scientific space missions, such as those for planet exploration. In the last few years, several instruments based on laser spectroscopy or interferometry have been launched on board satellites, planetary rovers, and other space vehicles, while other photonic instruments for new scientific mission are under development.

This chapter reviews the field of photonic sensors and instruments. The technology of sensors based on Fiber Bragg Gratings (FBGs) is introduced and the different approaches for angular velocity sensing through the lightwave technology are discussed. Finally, some examples of photonic instruments utilized in some recent space mission are reported.

6.1 Temperature and strain sensing by Fiber Bragg Gratings

The distributed monitoring of temperature and strain on a wide variety of sub-systems and structural panels of a spacecraft, during all their life cycle, is extremely useful to guarantee the success of a space mission [1]. The FBG is one of the most promising photonic technologies for stain/temperature mapping [2].

FBG sensors offer many advantages with respect to conventional technologies. In fact, they are characterized by insensitivity to electro-magnetic interference, light weight, low power consumption, high signal-to-noise ratio, and high accuracy. In addition, FBGs can be embedded within structures during their manufacturing. FBG technology has been used in Space, for the first time, when a multi-channel FBG system was utilized to monitor strains exerted on the liquid hydrogen fuel tank of the NASA DC-XA re-usable rocket (first flight in 1996) [3]. In the last two decades, it was experimented in several space applications, mainly related to the monitoring of the spacecraft propulsion subsystem, by NASA, ESA, and JAXA [4–7]. The performance of a complex sensor system based on FBG technology was evaluated in the space environment by the European Space Agency (ESA) mission Proba-2 [8, 9], in which a microsatellite was developed to demonstrate several innovative technologies in orbit, including FBG. The satellite, launched on November 2, 2009, included a sensor system based on 14 FBG sensors, which mainly monitors the Xenon propulsion subsystem consisting of the propellant tank, the supply assembly, and the thruster assembly. A pressure/temperature sensor monitors the Xenon propellant tank, a high-temperature sensor measures the transient temperature of the thruster, and the other 12 FBG sensors estimate the temperature in several strategic points of both the propulsion subsystem and other payloads.

An FBG is a periodic perturbation of the fiber core refractive index, which induces the reflection of light at a specific wavelength, which is called the Bragg wavelength (λ_B) [10]. Therefore, the FBG works as an optical filtering device that selects and reflects a certain portion of the input light spectrum. As already mentioned in Chapter 2, the Bragg wavelength is given by:

$$\lambda_B = 2n\Lambda, \tag{6.1}$$

where n is the effective refractive index of the optical mode propagating within the fiber and Λ is the grating period (or pitch). Strain and temperature change along the FBG induce variations in the effective refractive index and/or in the grating pitch, which in turn generate a shift in the Bragg wavelength.

The strain-induced change in the Bragg wavelength, which is linear with no evidence of hysteresis, is given by [10]:

$$\frac{\Delta\lambda_B}{\lambda_B} = \varepsilon_1 - \left(\frac{n^2}{2}\right)\left[p_{11}\varepsilon_t + p_{12}\left(\varepsilon_1 + \varepsilon_t\right)\right], \tag{6.2}$$

where ε_1 is the longitudinal strain along the fiber axis, ε_t is the strain transverse to the fiber axis, and p_{ij} are the elements of the stress-optic tensor. If the strain ε is homogeneous and isotropic, Eq. (6.2) can be written as:

$$\frac{\Delta\lambda_B}{\lambda_B} = \varepsilon\left(1 - p_e\right), \tag{6.3}$$

where all the photo-elastic contributions are included in p_e, which is defined as:

$$p_e = \left(\frac{n^2}{2}\right)\left[p_{12} - \mu\left(p_{11} + p_{12}\right)\right], \tag{6.4}$$

with μ being the Poisson ratio.

The temperature sensitivity of a FBG is mostly due to the thermo-optic effect. The shift in Bragg wavelength induced by temperature is given by [10]:

$$\frac{\Delta\lambda_B}{\lambda_B} = \left(\alpha_\Lambda + \alpha_n\right)\Delta T, \tag{6.5}$$

where α_Λ is the coefficient of thermal expansion of the fiber (about 0.55×10^{-6} °C^{-1} for a silica fiber), α_n is equal to $(1/n)$ (dn/dT), that is about 8.6×10^{-6} °C^{-1}, and ΔT is the temperature change. The FBG temperature response is not linear over a wide range and there is hysteresis. At the telecommunication wavelengths and near room temperature, λ_B varies with temperature by approximately 10 pm/°C.

In FBG sensors, the grating is excited either by a tunable laser or a broadband light source, and the Bragg wavelength shift is real-time monitored. Both the grating excitation and the Bragg wavelength monitoring are accomplished by a dedicated module, called an interrogator. The development of a space-borne interrogator module is reported in Ref. [11]. It is based on a modulated grating Y-branch (MGY) tunable laser diode, whose emission wavelength can be controlled electronically by tuning three currents. As shown in Fig. 6.1, two resonators are combined within the tunable laser source. One resonator is obtained through the mirror on the left of the device and grating 1, while the two reflecting elements forming the second resonator are the mirror on the left and grating 2. Both the gratings acting as reflectors are tuned by two control currents, I_{G1} and I_{G2}. The lasing wavelength depends on the overlap between the resonant modes of the two cavities.

The MGY laser can be tuned from 1,528 nm to 1,568 nm. During each measurement cycle, the laser scans all the tuning range and the reflected light from the FBG sensors located on the fiber connected to the laser is measured by a photodiode. In this way, the Bragg wavelengths of the FBGs are monitored.

One of the main advantages of FBG sensors is the multiplexing operation mode, i.e. several sensors with different grating periods can be

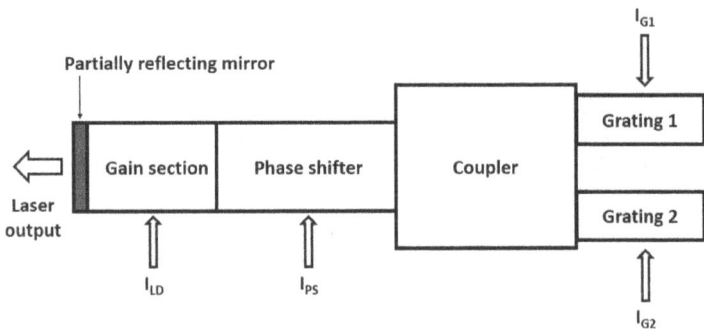

Fig. 6.1.   MGY tunable laser diode. The tuning currents are I_{PS}, I_{G1}, I_{G2}. I_{LD} is the current supplying the gain section.

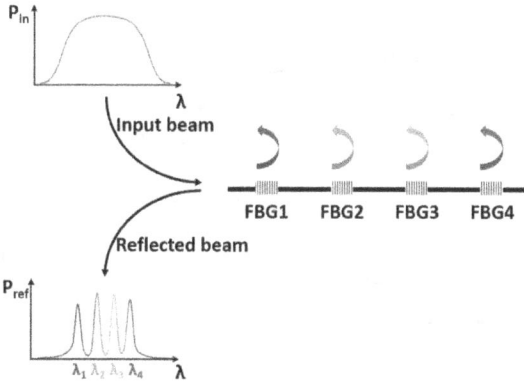

Fig. 6.2.   FBG sensor array. Gratings with different Bragg wavelengths are written serially in an optical fiber. P_{in} is the optical input power and P_{ref} is the reflected optical power.

inscribed along the same fiber to sense temperature or strain at a number of points. As illustrated in Fig. 6.2, the so obtained FBGs reflect light at different wavelengths that depend on the period of each grating.

The dual sensitivity to temperature and strain of FBG sensors is a serious critical aspect of this technology. This issue can be addressed by locating two gratings in each point to be monitored with very different responses to strain and temperature [12]. By measuring the responses to strain and temperature of both gratings during the sensor calibration, it is possible to distinguish the temperature and the stain effect by real-time monitoring the drift of both the Bragg wavelengths. In fact, the shift of the Bragg wavelengths λ_1 and λ_2 of the two gratings can be written as:

$$\begin{bmatrix} \Delta\lambda_1 \\ \Delta\lambda_1 \end{bmatrix} = \begin{bmatrix} K_{\varepsilon 1} & K_{T1} \\ K_{\varepsilon 2} & K_{T1} \end{bmatrix} \begin{bmatrix} \varepsilon \\ T \end{bmatrix}, \tag{6.6}$$

where $K_{\varepsilon 1}$ and $K_{\varepsilon 2}$ are the responses to strain ε and K_{T1} and K_{T2} are the responses to temperature T. $K_{\varepsilon 1}$, $K_{\varepsilon 2}$, K_{T1}, and K_{T2} should be evaluated during the grating calibration. By measuring $\Delta\lambda_1$ and $\Delta\lambda_2$, T and ε can be derived by inverting Eq. (6.6).

The phase mask technique [13] is the most common approach for FBG writing, which is based on the UV photosensitivity of optical fibers. The setup for the grating manufacturing is shown in Fig. 6.3. A phase

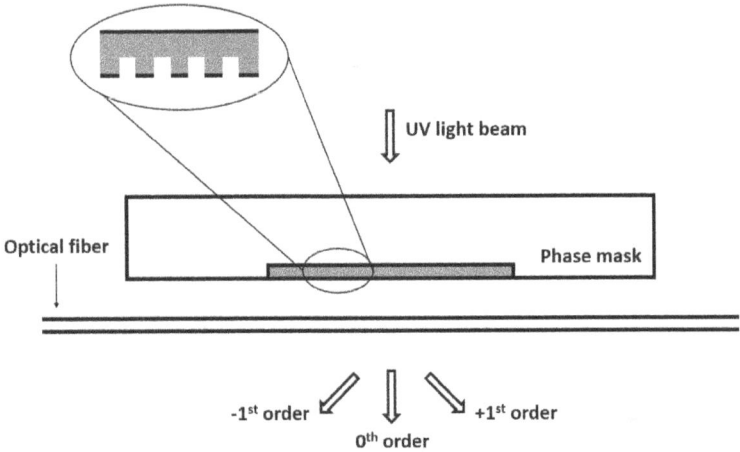

Fig. 6.3.   FBG fabrication by the phase mask technique.

mask, which is transparent to UV light, is placed between the UV light source (e.g. a KrF excimer laser) and the photosensitive fiber. The mask is a slab of silica glass on which a periodic structure is etched using a photolithographic technique. The UV beam passes through the phase mask and is diffracted by the periodic corrugations of the phase mask, which is designed to suppress the zero-order diffracted beam. Most of the diffracted optical power is relevant to the ±1 diffracted orders, which interfere to produce a periodic pattern that photo-imprints the grating on the optical fiber. The period of the grating obtained onto the fiber is independent of the wavelength of UV light irradiating the mask, and it is equal to $\Lambda_{mask}/2$, with being Λ_{mask} the period of the phase mask grating. Conversely, the corrugation depth required to suppress the zero-order diffracted beam depends on the wavelength of the UV beam.

Another manufacturing process for FBGs is grating writing by femto-second infrared (fs-IR) pulses [14, 15]. By using this technique, the FBG formation is due to a nonlinear multiphoton absorption/ionization process leading to the structural modification of the fiber core, and the grating can be manufactured either by using a phase mask or by direct writing.

The study of effects of radiation on the FBG is obviously crucial for the application of this technology in the space environment. When an

FBG written into an optical fiber is irradiated with gamma rays, that irradiation can induce changes in the effective refractive index (Δn_{eff}) and/or in the period ($\Delta\Lambda$). Both Δn_{eff} and $\Delta\Lambda$ induce Bragg wavelength shifts.

The behavior of the Bragg wavelength under γ-radiation depends on the fiber chemical composition and the conditions of the grating writing. Generally, γ-irradiated FBGs show a red shift of Bragg wavelength and a saturation effect at specific values of the dose. For example, the effect of γ-radiation on a 10 mol% Ge-doped silica fiber is a red-shift of the Bragg wavelength up to 20 pm for a dose of 100 kGy [16]. That shift does not increase when the dose is increased up to 1 MGy. The nitrogen-doped silica fibers do not exhibit the saturation of the Bragg wavelength shift, which increases quite linearly at least up to a 1.5 MGy dose [16]. To enhance the UV photosensitivity of optical fibers, they can be put into a high-pressure vessel containing hydrogen before the grating writing. This technique, which is named hydrogen loading, increases the red shift of the Bragg wavelength due to γ-radiation. When a hydrogen loaded Ge-doped fiber is irradiated with γ rays at a total dose of 100 KGy, the red-shift of the Bragg wavelength is up to 60 pm [16]. The Bragg wavelength shift due to γ-radiation is low, i.e. 3–7 pm with total dose = 100 kGy, if the FBG is written in fluorine-doped fibers using the fs-IR technology [17].

The influence of the grating temperature during the γ-irradiation has been investigated in Ref. [18]. The Bragg wavelength shift at +78°C is approximately 30% lower than that at room temperature, whereas at −50°C the radiation-induced Bragg wavelength shift is about two times larger than at room temperature.

Another factor that influences the FBGs response to γ-radiation is the polymer coating. In fact, γ-radiation can cause the swelling of polymer coatings, with a consequent shift of the Bragg wavelength [19]. Thus, stripped fibers have a lower sensitivity to radiation than polymer coated ones.

6.2 Photonic gyroscopes

All spacecrafts include an attitude and orbit control sub-system (AOCS) that senses the craft attitude, compares it to the desired attitude, and generates the commands for the actuators, which actively change the attitude by creating torque [20]. The attitude sensing functionality can be carried out by several sensors, which measure both the absolute attitude and the spacecraft angular rate in an inertial reference frame. The former estimation is accomplished by star trackers, sun sensors, earth sensors, magnetometers, and so on, while gyroscopes (or gyros) perform the latter.

The gyroscope has a large variety of space applications, which impose different requirements on the sensor. The main gyro performance parameters, which are utilized to specify the application requirements, are resolution, i.e. minimum detectable angular velocity, bias drift (the bias is the non-zero output when the sensor does not rotate), and angle random walk (ARW), which is a noise contribution with a variance linearly increasing as time increases, observed when the gyro is utilized to measure the rotation angle by integrating the sensor output.

Although progress in star tracker technology (see the previous chapter) currently allows the development of gyro-less satellites, e.g. Iridium Next and Globstar 2 constellations [21], gyroscopes with ultra-high resolution (<0.1°/h) are still essential in Earth observation and scientific missions. Angular velocity sensors with resolution of the order of 1–10°/h are useful in all satellites and rover vehicles for planet exploration.

For at least two decades, photonic sensors have dominated the market of gyros for space applications. These devices are all based on the Sagnac effect [22], i.e. the generation of a phase (or frequency) shift between two optical beams that counter-propagate along a rotating closed path.

The first photonic gyroscope for space applications was the ring laser gyro (RLG), which is still widely used, also on board launchers. The fiber optic gyro (FOG), which exhibits a bias drift ranging from 0.1°/h to 0.0003°/h, has been selected in many space missions, especially in Europe, and was mounted on board the planetary rovers for Mars exploration developed by NASA [23]. Until now, the only two photonic gyroscopes, which are available on the market, are the FOG and the RLG.

In recent years, the research effort aimed at the miniaturization of photonic gyros for space applications and lowering their cost has been constantly increasing. Several approaches, based on either semiconductor ring lasers or integrated passive cavities excited by external laser sources, have been explored, but currently the performance of the best performing prototypes is still far from that required by space applications.

The market demand for low-cost miniaturized gyroscopes in Space is quickly growing and the MEMS gyros for space, e.g. those developed under some ESA contracts [24], still seem to be unable to fully satisfy the demand of that market. This means that an enhancement of the research effort on miniaturized photonic gyros is very useful because it will try to respond to market demand, which is still not fully covered.

6.2.1 *He–Ne ring laser gyro*

The first He–Ne RLG was demonstrated in 1963 [25], but only entered the market in the 1980s, replacing mechanical gyroscopes, thanks to its improved lifetime and reliability. Currently, the RLG is a key sensor for the aerospace and defense industry, and is widely used in many application domains requiring high-performance gyros, such as strapdown inertial navigation and missile guidance [26].

The success of the He–Ne RLG is due to a great number of advantages when compared to mechanical gyros, such as its insensitivity to vibrations and temperature gradients, its digital output, the absence of moving parts, the design simplicity (less than 20 components), wide dynamic range, fast update, and good reliability. These features allow high performance, i.e. resolution below 0.1°/h and bias drift better than 0.01°/h [27].

Considering the possible future trends of the space industry, the main drawbacks of this device are size, mass, power consumption, high cost, and possible degradation of the He–Ne mixture, which is responsible for the optical gain within the laser resonant path. To overcome the latter critical aspect, the replacement of the He–Ne gain medium with a solid-state one, Nd:YAG, has been proposed [28].

In the RLG, two counter-propagating resonant modes are excited within an optical cavity where a gain medium generates optical amplification of passing light. Due to the Sagnac effect, the difference between the frequencies of the two optical modes propagating within the cavity is proportional to the sensor angular rate Ω. The RLG basic configuration is show in Fig. 6.4. The cavity shape can be either triangular or square.

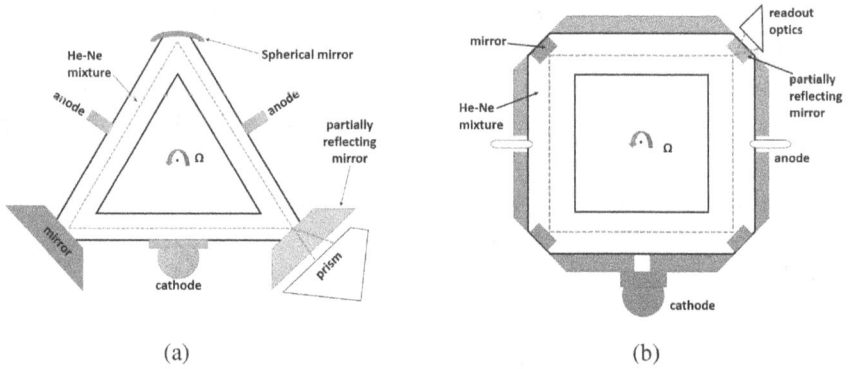

Fig. 6.4.   (a) He–Ne RLG based on a triangle optical cavity. (b) He–Ne RLG based on a squared optical cavity.

The former includes two corner mirrors and a spherical mirror, while in the latter, four corner mirrors are used. The counter-propagating laser beams interfere at the partially reflecting mirror and produce a fringe pattern, which shifts when a rotation rate is applied along an axis perpendicular to the active cavity plane [29].

A high voltage applied across the electrodes (anodes and cathode) is used to ionize the He–Ne gas. Ionization excites Ne atoms which radiatively decay to lower energy levels, generate the lasing beams. Mode competition between the oppositely directed waves must be avoided in order to ensure the same optical power level to the modes propagating in the active cavity. Moreover, to avoid thermal effects and enhance mechanical stability, He–Ne RLGs are usually fabricated within a solid block of a material with a very low coefficient of thermal expansion (e.g. vitroceramic glass).

The frequency difference between the lasing modes is given by [29]:

$$\Delta v = \frac{4A\Omega}{\lambda_0 P} = S\Omega, \qquad (6.7)$$

where A and P are the laser cavity area and perimeter, respectively, λ_0 is the lasing wavelength and $S = (4A\Omega)/(\lambda_0 P)$ is the sensor scale factor.

The readout optics utilized to measure the frequency difference Δv usually consists of a prism combined with a photodetector array. The two laser beams interfere to form a fringe pattern at the output of the prism. The detector array reads the time-dependent intensity of light on its surface, which is given by [27]:

$$I(x,t) = I_0 \left[1 + \cos\left(2\pi\Delta v t + \frac{2\pi}{\lambda}\gamma_p x + \varphi_0 \right) \right],\qquad(6.8)$$

where I_0 is the average value of $I(x,t)$, γ_p is the angle between the beams at the output of the prism, φ_0 is a constant phase difference between the beams, x is the spatial coordinate measured along the detector array, and t is the time instant. During the integration time $\Delta\tau_i$, the detector array is capable of reading a number N_m of intensity maxima which is given by:

$$N_m = \int_0^{\Delta\tau_i} \Delta v \, dt = S \int_0^{\Delta\tau_i} \Omega \, dt = S\theta,\qquad(6.9)$$

where θ is the gyro rotation angle during the time interval $\Delta\tau_i$. There is a linear relationship between the rotation angle and the number of fringes counted by the detector.

The three main noise sources in He–Ne RLG are bias drift, mode locking, and scale factor variations.

The first one is relevant to a non-zero frequency difference between the lasing beams for $\Omega = 0$, which implies a shift of the sensor input/output ideal characteristic by an amount equal to K_0, corresponding to the bias. Currently, the typical bias drift of He–Ne RLGs is in the range 0.005–0.01°/h.

Mode locking (or lock-in) leads to a nonlinear relation between the frequency difference Δv and the rotation rate Ω.

The ideal and the typical input/output characteristics of the RLG are shown in Fig. 6.5. When the angular rate is less than a critical value $|\Omega_L|$, the two waves propagating in opposite directions lock and the frequency difference between the lasing modes is null, i.e. $\Delta v = 0$. The RLG is insensitive to rotation rate in this dead band extending from $-\Omega_L$ to $+\Omega_L$ (Ω_L is the dead band edge).

Mode locking is mainly due to the coupling between counter-propagating waves and backscattered waves. To reduce backscattering, related to imperfections in mirror or other intra-cavity optical elements,

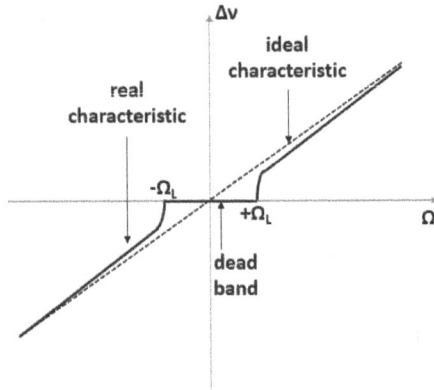

Fig. 6.5.   Mode locking effect in the He–Ne RLG.

high accuracy for mirror manufacturing and polishing is required. This makes He–Ne RLG fabrication very expensive.

The most common countermeasure against the lock-in is the application to the sensor of a time-varying mechanical rotation, called dithering, controlled by a sine-wave signal. The rotation is imposed through a piezoelectric transducer, while the initial phase relevant to every dither cycle is randomly varied [30]. A ring cavity supporting more than one pair of counter-propagating modes can be used to significantly reduce the lock-in effect. In Ref. 31, an RLG is proposed with two resonant modes propagating in the clockwise direction and the other two propagating in the opposite direction. One of the clockwise beams is right-circularly polarized (RCP) and the other one is left-circularly polarized (LCP). The beams in the opposite direction have the same polarizations. A reciprocal bias produced by a polarization rotator or a non-planar cavity is used to split the frequency of co-propagating modes having different circular polarization, whereas a non-reciprocal bias achieved by the Faraday effect splits the resonant frequencies of counter-traveling modes having the same polarization. In this sensor, the angular velocity is proportional to the frequency difference between the LCP and the RCP resonant modes.

Due to scale factor variations, the nominal value of S is not equal to the experimental value. According to Ref. [32], the experimentally measured value of the gyro scale factor is equal to $S(1-\delta)$, where δ is a coefficient depending on the ring laser output power, the laser gain, and the gain medium dispersion.

6.2.2 *Interferometric fiber optic gyro*

In the early 1970s, the development of laser diodes and low-loss single-mode optical fiber allowed FOG to be developed as practical devices [33]. The investigation of optical angular rate sensors based on optical fibers was aimed at reducing cost, simplifying the fabrication process, and increasing accuracy with respect to the He–Ne RLGs. Several gyro configurations based on different kinds of optical fibers, e.g. polarization maintaining, hollow core, and erbium doped fibers, have been conceived and investigated in the last decades, but only the interferometric FOG (IFOG) has reached the commercialization stage.

The basic IFOG configuration is shown in Fig. 6.6. A beam splitter is used to split the light from a light source into two different counter-propagating beams. The beams are coupled into the two ends of a multi-turn fiber coil by two lens systems. After propagating within the fiber coil, the counter-propagating waves are recombined by the beam splitter and the optical signal resulting from the interference is sent to the photodetector.

When the sensor rotates around the axis orthogonal to the coil plane, a phase difference $\Delta\varphi$ between the two counter-propagating optical beams is induced by the Sagnac effect. The phase difference is proportional to the angular rate Ω and it is given by:

$$\Delta\varphi = \frac{8\pi^2 R^2}{c\lambda_0} k\Omega = \frac{4\pi LR}{c\lambda_0}\Omega, \qquad (6.10)$$

where λ_0 is the sensor operating wavelength, R is the fiber coil radius, c is the speed of light in the vacuum, k is the number of coil turns, and L is the fiber total length.

Since the optical path directly influences the gyro scale factor, it is possible to enhance the gyro sensitivity by increasing k. Unfortunately, because of the fiber attenuation, the fiber length cannot increase indefinitely.

At rest, since the two beams at the PD are not in phase, the basic IFOG configuration in Fig. 6.6 induces a phase shift of $\pi/2$ between transmitted and reflected beams. To overcome this critical aspect, the so-called reciprocal (or minimum) configuration shown in Fig. 6.7 is used. In the reciprocal configuration, the CW and CCW signals are both reflected and transmitted passing through the beam splitters two times, so that the

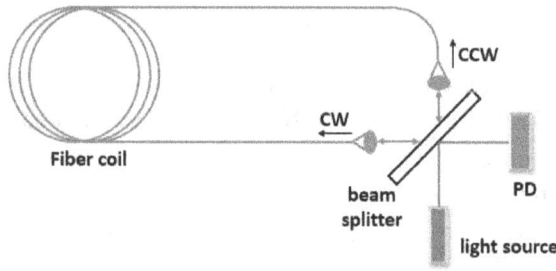

Fig. 6.6.   Basic configuration of the IFOG. CW: clockwise. CCW: counter-clockwise. PD: photodetector.

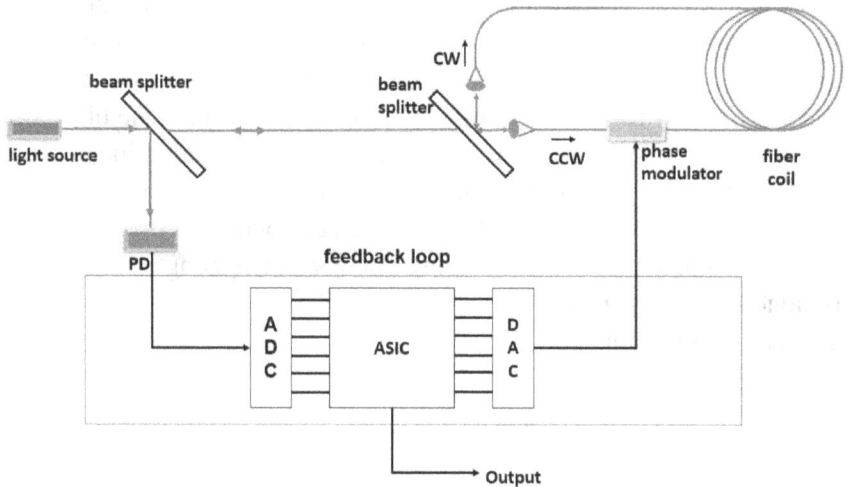

Fig. 6.7.   Closed loop minimum IFOG configuration.

beams at the input of PD are in phase when the gyro is at the rest [34]. By inserting a phase modulator after the beam splitter, it is possible to keep the gyro operating at the point where the sensitivity to rotation is maximum [35]. The PD and the phase modulator are connected by a feedback loop, which nulls the phase shift due to the Sagnac effect by imposing a controlled phase shift on the CCW beam, which is proportional to the sensor angular velocity.

The feedback loop includes an analog-to-digital converter (ADC), an application-specific integrated circuit (ASIC) and a digital-to-analog converter (DAC). The ADC receives and converts into electrical form the analog output signal of the PD. The ASIC processes the electric signal, in order to generate the digital form of modulating signal and the sensor output. Finally, the DAC converts the digital form of the modulating signal into the analogue one.

Since IFOG performance can be enhanced by increasing the length of the fiber coil and/or properly designing the readout opto-electronic system, this gyro has the potential to achieve very good performance, better than that of the RLG. This can be obtained if all the sources of non-reciprocity, i.e. Kerr nonlinearity of silica, time varying temperature fluctuations along the fiber coil (Shupe effect), vibrations of the fiber coil, polarizations instabilities and Rayleigh backscattering, are properly minimized.

Rayleigh backscattering can be minimized by using a broadband (emission spectrum width around 30–50 nm) light source because its very short coherence length allows the suppression of the spurious interference effects involving the waves generated by back-scattering/back-reflections and inaccuracies in the components controlling the polarization of the beams propagating in the fiber coil. Two different broadband sources are utilized, super luminescent diodes (emitting at 800 nm) or sources based on amplified spontaneous emission in erbium-doped fibers (emitting at 1,500 nm). Broadband sources based on erbium-doped fibers are preferred in high-performance IFOGs thanks to their high stability with respect to temperature drift.

The Shupe effect can be minimized by particular winding schemes, such as the quadrupolar one [36], utilized in the coil fabrication. The best countermeasure against the polarization noise is the use of polarization maintaining fibers.

6.2.3 *Space applications of the He–Ne RLGs and the IFOGs*

As already mentioned, the performance of both RLGs and FOGs makes them suitable in a variety of space missions. In Table 6.1, the features of some RLGs and FOGs already used in space are summarized.

The RGA-14 RLG has been selected to operate on the 66-satellite IRIDIUM constellation [37], which includes telecommunications

Table 6.1.   Features of some RLGs and FOGs already used in space [36–40].

	ASTRIX 200	ASTRIX 1120	ASTRIX 120	RGA-14	RGA-20	CIRUS	LN-200N	MIMU	FOG-based IRS
Company	Airbus Def. & Space	Airbus Def. & Space	Airbus Def. & Space	L3 Comm.	L3 Comm.	L3 Comm.	Northrop Grumman	Honeywell	Honeywell
Technology	FOG	FOG	FOG	RLG	RLG	FOG	FOG	RLG	FOG
Bias drift (°/h)	0.0005	<0.003	0.01	0.05–0.2	0.005–0.01	0.0003	0.1	<0.005	<0.0003
ARW (°/√h)	0.00012	0.002	0.0016	0.005–0.03	0.0025–0.004	0.00019	0.07	<0.005	<0.0001
Power cons. (W)	24	12	24	<16	<25	<25	12	<32	—
Volume (cm³)	<8,000	1,500	1,200	3,800	7,800	2,500	240	1,200	—
Mass (kg)	13	4.2	6.7	2.2	5.27	13.5	0.75	4.7	—

satellites in low Earth orbit, while the RGA-20 RLG has been used on several satellites, including IKONOS, a commercial Earth observation satellite, QuikSCAT, a NASA satellite measuring the surface wind speed and direction over the ice-free global oceans, FUSE, a NASA satellite observing the universe using high resolution far-UV spectroscopy, and MTI, a US satellite for multispectral imaging [38].

The development of the ASTRIX family [39] of FOGs by Astrium and Ixsea was supported by both ESA and CNES (the French *Centre National d'Études Spatiales*). Currently, these gyros are used in spacecrafts.

The miniature inertial measurement unit (MIMU) [40], based on RLG technology, has been widely utilized on board of aircraft and has been selected fort many Space missions, e.g. the NASA missions to Mars, Rosetta and Mars Express, and to Venus, Venus Express.

The LN-200 FOG-based inertial measurement unit [41] has been used in some spacecraft and enabled the autonomous navigation of the NASA planetary rovers.

6.2.4 *Resonant fiber optic gyro (RFOG)*

RFOG [42] could potentially achieve IFOG performance with a coil length up to 100 times shorter than those of IFOGs, but it is still far from the commercialization stage, mainly because it has not yet exhibited any clear and significant advantage over the IFOGs in terms of cost, size, weight, power consumption, or reliability.

The sensing element of the RFOG consists of a fiber ring resonator formed by a single-mode fiber, which is evanescently coupled to one or two fibers for the cavity excitation. The resonator, when properly excited, supports two counter-propagating resonant modes, which have the same resonance frequency when the sensor does not rotate. When $\Omega \neq 0$, the resonance frequencies are different and, according to the Sagnac effect, their difference $\Delta\nu$ is proportional to Ω.

The resonator configuration including the ring coupled to just one fiber, which is the most common configuration, is shown in Fig. 6.8. Since the RFOG operation requires the simultaneous excitation of two resonant modes within the cavity, the exciting beams are launched at the two ends of the fiber coupled to the resonator. Two isolators, not shown in Fig. 6.8, are used to separate the input and the output at each fiber end.

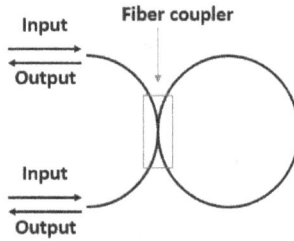

Fig. 6.8.   Typical configuration of the fiber ring resonator with one fiber coupler.

The resonance condition for the fiber ring resonator is given by:

$$\beta \pi d = 2q\pi, \tag{6.11}$$

where q is an integer number called the resonance order, β is the propagation constant within the ring and d is the ring diameter.

As already mentioned, when the resonator is at rest, the resonance frequencies of both optical modes propagating within the resonator coincide, while when the sensor rotates the frequencies exhibit a difference Δv, which can be written as:

$$\Delta v = d \frac{v_0}{c} \Omega, \tag{6.12}$$

where v_0 is the cavity resonance frequency when $\Omega = 0$.

It can be observed that the frequency difference Δv is directly proportional to the angular rate of the cavity. If a multi-turn fiber coil is used to manufacture the resonator, the turn number does not affect Δv, which only depends on the coil diameter and the sensor operating frequency [43]. The turn number can be optimized to optimize the cavity Q-factor, which should be enhanced to improve the gyro resolution.

The main technical challenge for the RFOG is the precise determination of Δv. The typical configuration of the sensor consists of the fiber resonator, which acts as sensing element, an optical system enabling the resonator excitation, and a readout opto-electronic circuit allowing estimating Δv. Three options are available for the RFOG readout system: phase or frequency modulation of the optical signals exciting the fiber

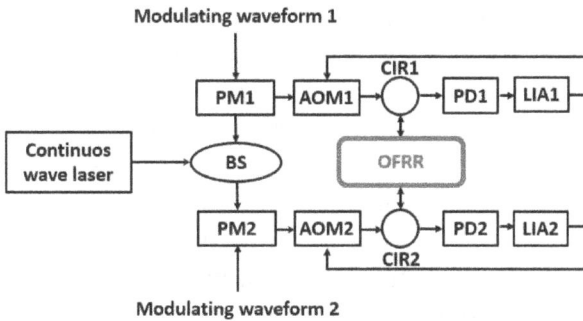

Fig. 6.9.   RFOG readout system based on the phase modulation spectroscopy.

resonator, and modulation of the fiber resonator length by a piezoelectric transducer.

The RFOG configuration using the phase modulation spectroscopy technique for the readout circuit [44] is shown in Fig. 6.9.

The optical fiber ring resonator (OFRR) is the element sensitive to rotation. A 50/50 beam splitter (BS) splits the laser beam and the resulting beams are phase modulated by two phase modulators PM1 and PM2, driven by sinusoidal waveforms. After the phase modulation, the optical beams are frequency shifted by two acousto-optic modulators, AOM1 and AOM2, and excite the fiber ring resonator. The signals at the output of the OFRR are sent to the photodetectors PD1 and PD2, which transduce the optical signals into electric ones. The OFRR is excited by two circulators (CIR) because we assume that the ring is coupled to just one fiber (resonator configuration in Fig. 6.8).

The electric signals coming out from PD1 and PD2 are demodulated by the lock-in amplifiers LIA1 and LIA2, which generate the error signals for the two feedback loops. The loops lock the central frequency of the beams, thus exciting the cavity to the two resonance frequencies of the OFRR. Since the difference Δv between these two frequencies is proportional to Ω, the difference between the feedback electric signals driving AOM1 and AOM2 is proportional to Ω.

The ultimate limit to RFOG resolution is imposed by the shot noise of the photodetectors. The other sensor noise sources are Raylegh backscattering [45], polarization fluctuation [46, 47], and the Kerr effect within the OFRR [48].

Unlike IFOGs, it is not possible to use a broadband light source to reduce the influence of those noise sources, which can be mitigated by using several approaches. The most common technique to reduce the influence of the backscattering is carrier suppression in the beams exciting the OFRR through the appropriate selection of the amplitude of the modulating waveforms. One way to limit the inaccuracy due to the Kerr effect is to reduce the optical power of the resonating modes propagating within the fiber resonator. As in IFOGs, polarization noise can be reduced by using polarization maintaining fibers.

Recently, the potential advantages of air-core photonic band gap fibers in the RMOG technology has been explored. These fibers, consisting of a hollow core and a cladding made of fused silica incorporating an array of air holes, are very immune to the Kerr effect and thermally driven polarization instabilities. An RFOG with the OFRR made of an air-core photonic band gap fiber has been demonstrated [49], even if the performance in terms of bias drift (1,800°/h) is still far from that of the IFOG.

To our knowledge, the best performing RFOGs [50] exhibit a resolution of the order of 1°/h, about two orders of magnitude worse than that of the best performing IFOGs, which are commercial off-the shelf products, specifically targeted for Space.

6.2.5 *Integrated optical gyroscope*

As previously mentioned, miniaturization and cost lowering are currently the main challenges for gyro technology, especially in the Space context. The use of integrated optics to develop new gyros is considered one of the most promising ways to face these challenges [51]. In fact, the use of integrated optics allows the fabrication of a number of integrated optical circuits/microsystems on a single chip, namely photonic integrated circuits (PICs), and considerable research effort has been already spent on designing and fabricating PICs for various applications. In the development of new gyroscopes for space applications, integrated optical technology is very attractive because it potentially enables weight and size, lowered cost, decreasing power consumption, better control of the thermal effects, increased reliability, and very good resistance to the harsh environment.

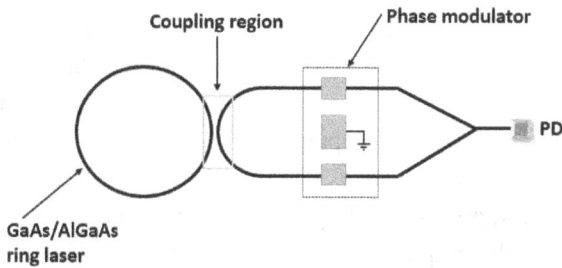

Fig. 6.10. Configuration of the SRL-based gyro in Ref. [53].

The vast majority of the research activity on the integrated optical gyro is focused on the sensors based on semiconductor ring lasers (SRLs) and resonant micro optic gyros (RMOGs). In the former kind of device, the sensing element is a ring laser generating two counter-propagating lasing modes, which experience a rotation-induced frequency shift that can be measured by an interferometric technique [52]. The RMOG basic configuration includes a passive ring resonator with high Q-factor, a narrow-linewidth laser, and other optical/opto-electronic components for the sensor readout. The operating principle of the SRL-based gyro is similar to that of the He–Ne RLG, while the RMOG has the same operating principle as the RFOG.

The configuration of a patented configuration of an SRL-based gyro [53] is shown in Fig. 6.10. The main gyro components are a GaAs/AlGaAs double quantum well circular SRL, which is the element sensitive to rotation, a circular directional coupler, an electro-optic phase modulator, a Y-junction, and a photodetector. When the sensor rotates around the axis orthogonal to the substrate, the two beams generated by the ring laser have frequencies which are slightly different. According to the Sagnac effect, the difference Δv between the frequencies is proportional to Ω. The laser-generated beams interfere in the Y-junction and a sine-wave electric signal with frequency Δv (beating signal) is generated at the photodiode output. The phase modulator, by imposing a constant phase shift of $\pi/2$, allows the derivation of the sense of rotation.

Although, in the last decades, different configurations of SRL-based gyros have been both theoretically and experimentally investigated, no satisfactory experimental results in terms of rotation rate have been reported in the literature. The main critical aspects of SRL-based gyros are the lock-in, due to the backscattering within the ring laser, and the mode competition in the active medium, which can be responsible for preventing the presence of two counter-propagating waves. Only the identification of the effective countermeasures for those effects could allow the demonstration of the first prototype of an SRL-based gyro having a performance compliant with the requirements imposed by the different application domains.

To overcome the critical effects of SRL-based gyros, the laser source and the sensing element can be decoupled, as in the RMOG, where a passive ring resonator [54] is the element sensitive to rotation and an external laser source generates the counter-propagating beams exciting the cavity [55].

The typical configuration of the RMOG, which is conceptually similar to that of the RFOG, consists of a waveguide ring resonator acting as sensing element, several optical and opto-electronic devices for cavity excitation and the sensor readout, two photodiodes, and an electronic readout unit (see Fig. 6.11).

When the device is at rest, the resonance frequencies of the two counter-propagating modes in the ring resonator are equal. When the device rotates, the angular rate is measured by the difference between the

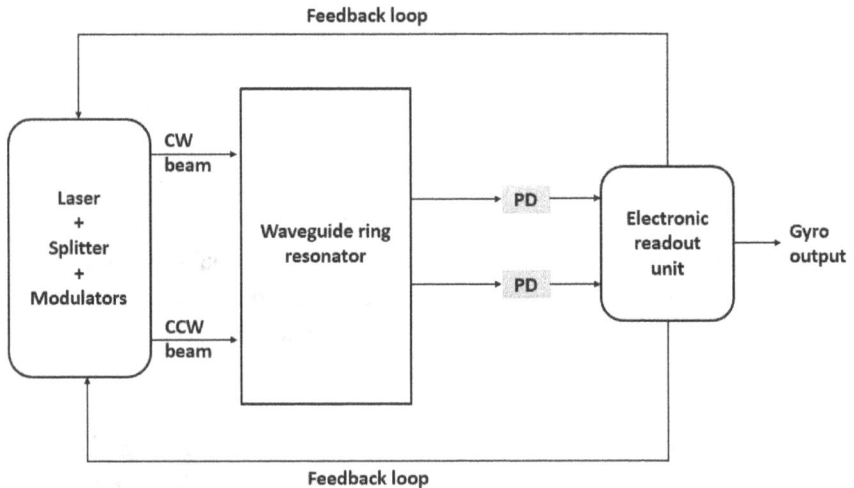

Fig. 6.11.   Basic configuration of a RMOG.

resonance frequencies of the two counter-propagating modes injected into the ring resonator. The technique for the measurement of this difference is the same already discussed for the RFOG. Basically, the functionality of the two feedback loops is to lock the central frequency of the beams exciting the resonator to the resonance frequencies of the cavity. Several configurations of readout circuits properly designed for RMOGs have been reported in the last few years [56–58].

Independent of the configuration of the readout system, the shot noise at the photodetectors imposes the ultimate limit on the gyro resolution. The shot noise limited resolution of the RMOG is equal to [59]:

$$\delta\Omega = \frac{1}{Qd\sqrt{P_{PD}}}\sqrt{\frac{2hc^3}{\lambda_0\eta\tau_{int}}}, \tag{6.13}$$

where d is the diameter of the ring resonator, Q is its quality factor, λ_0 is the sensor operating wavelength, τ_{int} is the sensor integration time, P_{PD} is the average power at the photodetectors, η is the photodetector quantum efficiency, h is the Planck's constant, and c is the speed of light in vacuum. The sensor resolution strongly depends on the resonator Q-factor, that should be increased by lowering the propagation loss of the waveguide forming the cavity, hence the design of the ring resonator cavity is crucial in RMOGs since it strongly affects the gyro minimum detectable angular velocity. To maximize the gyro performance, different material systems for the fabrication of the ring resonator have been considered. In Refs. [60–62], RMOGs based on silica-on-silicon ring cavities with a high Q-factor ($>10^6$) have been reported. Recently, some numerical results have suggested that silica-on-silicon 1D PhC ring resonators, consisting of ring resonators with a Bragg grating included in the resonant path, could have a Q-factor exceeding one billion [63].

One of the most interesting features of the RMOG is the potential integrability of all optical/opto-electronic sensor components on a single chip. In an ESA-funded project, the feasibility of a gyro-on-chip with target resolution of 10°/h, based on the RMOG configuration, has been theoretically and experimentally investigated [64, 65]. InP technology, due to its maturity and versatility, was selected for the sensor manufacturing. The InP gyro-on-chip has been designed and an experimental demonstration

including only the InP ring resonator (radius = 13 mm) evanescently coupled to one bus waveguide has been given [66]. The performance of the
cavity in terms of Q-factor = 10^6, finesse = 5.3, waveguide propagation loss
= 0.45 dB/cm, and resonance depth = 7 dB, are very encouraging [67]. The
functionality of the RMOG including the InP cavity, a narrow-linewidth
fiber laser, other optical/opto-electronic components and a custom readout
electronic board has been demonstrated in Refs. [68–71].

While shot noise represents an unavoidable noise in the RMOGs, the
effective mitigation of the three major error sources, backscattering-
induced noise, polarization noise, and Kerr-effect, still represents a critical
aspect to be overcome.

The best value of long-term bias stability for the RMOG technology
achieved so far is about 300°/h [72]. This performance does not match the
requirements of any space applications. Thus, an improvement of the
RMOG performance of two orders of magnitude is still necessary for the
RMOG to be used in spacecraft design. In addition, the RMOG prototypes
proposed in the literature are bulk systems including packaged and fiber
pigtailed optical-opto-electronic components connected through optical
fiber cables. A notable R&D effort is needful to make these sensors attractive in terms of mass, volume, and power consumption. Currently the
Technology Readiness Level (TRL) of the RMOGs for space is 3–4.

In recent years, some research groups have envisaged that multi-ring
optical structures, such as coupled resonator optical waveguides (CROWs),
can be used as fundamental building blocks for the development of ultra-
small gyros [73–76]. The research activity on this topic is currently in an
early stage and there is no experimental evidence on the performance
enhancement provided by multi-rings optical structures. In Ref. [77], the
authors claim, based on their theoretical model, that a properly optimized
CROW including 21 resonators and having a total footprint < 1mm^2 can be
used as the sensing element of a gyro with a shot noise limited minimum
detectable rotation rate of 0.002°/h.

6.3 *In situ* analysis instruments

In several space missions, rover vehicles, moving across the surface of a
planet or other celestial body, are utilized to make observations at

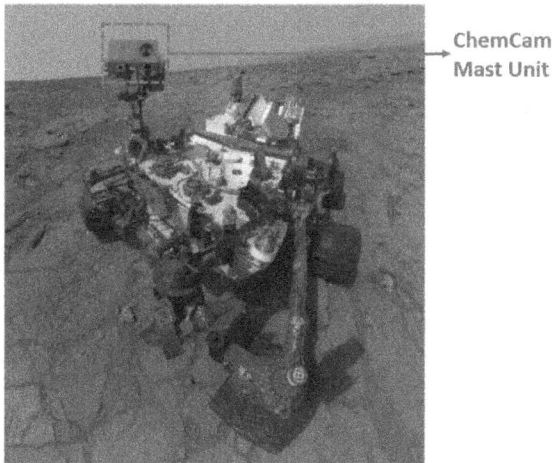

Fig. 6.12.   Picture of the rover Curiosity.
Source: Adapted with permission from Fig. 2 of Ref. [81].

microscopic level and to conduct experiments [78]. To perform these activities, the rovers are equipped with several instruments. One of the most interesting classes of such instruments is that based on laser spectroscopy [79], which, generally speaking, studies emission and absorption of light by matter.

The ChemCam instrument [80], mounted on board the NASA rover Curiosity (Fig. 6.12), which completed a Martian year on June 24, 2014, provides one of the most successful examples of laser spectroscopy capability in *in situ* planetary exploration. ChemCam has already demonstrated its ability to determine the elemental compositions of rocks and soils up to 7 m away from the rover.

The operating principle of ChemCam is laser-induced breakdown spectroscopy (LIBS) [82]. In LIBS instruments, a pulsed laser with irradiance on the target >10 MW/mm^2 induces the localized vaporization of the target and the generation of a transient plasma. Atoms and ions in the plasma are excited and emit their characteristic spectral lines. By analyzing the spectrum of the so generated radiation, qualitative and quantitative information on the target, i.e. its chemical composition, can be derived.

The block diagram of ChemCam, having an overall mass of 5.62 kg, is shown in Fig. 6.13. The instrument consists of two units, i.e. the mast

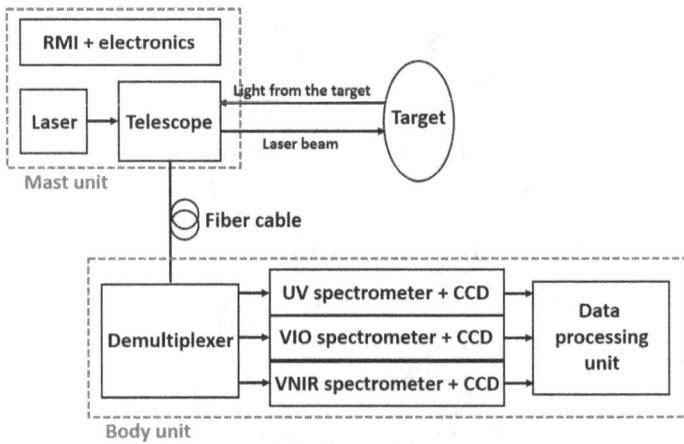

Fig. 6.13.   Block diagram of ChemCam. CCD: Charge-coupled device, UV: Ultraviolet, VIO: Violet, VNIR: Visible and near infrared.

unit (see Fig. 6.12), included in the rover mast, and the body unit. The pulsed laser beam is generated in the mast unit while the body unity allows the processing of the light generated by the vaporization of the target and the consequent generation of plasma. A telescope is located in the mast unit to collect the light coming from the target. The Remote Microimager (RMI) with its front-end electronics, both included in the mast unit, provides context imaging of the LIBS observation points, which is useful to quantify how the surface morphology influences the LIBS spectra.

The key elements of the body unit are an optical demultiplexer, three spectrometers each covering a specific spectral range, three CCD-based detector arrays with their thermoelectric coolers, and the data processing unit. The mast and the body unit are connected by a fiber optic cable (length about 6 m), which transfers the light collected from the telescope to the demultiplexer.

The laser emits optical pulses having energy = 20–35 mJ and duration = 5–8 ns, at 1,067 nm, with a repetition rate of 1–10 Hz [83]. The optical gain is provided by three Potassium–Gadolinium–Tungsten crystals doped with Neodymium (Nd:KGW), which are quite insensitive to the thermal shift of the pumping wavelength because of its very wide absorption linewidth [83]. Each crystal is optically pumped by a 700 W diode stack.

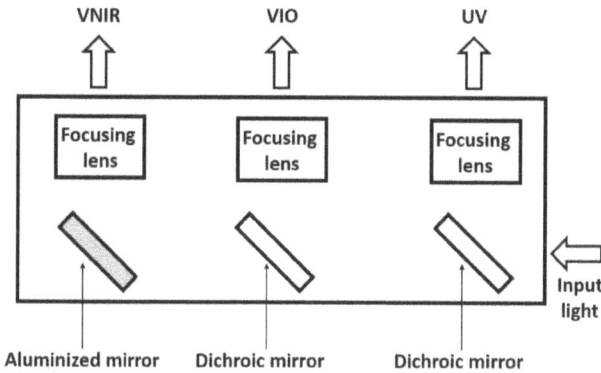

Fig. 6.14. Configuration of the demultiplexer included in ChemCam.

The pulse operation is obtained by Q-switching. The laser (weight about 0.6 kg) has a cylindrical shape and its dimensions are: diameter = 5.5 cm and length = 22 cm.

The three spectrometers operate in different wavelength ranges, from 240.1 nm to 342.2 nm (ultraviolet, UV), from 382.1 nm to 469.3 nm (violet, VIO), and from 474.0 nm to 906.5 nm (visible and near infrared, VNIR). The optical demultiplexer, having one input and three outputs, as in Fig. 6.14, divides the light coming out from the fiber cable into three beams. The beam at each demultiplexer output only has spectral components in one of the three bands, i.e. UV, VIO, or VNIR.

The spectrometers, which are connected to the demultiplexer outputs, spatially separate the spectral components of the incoming light, allowing the estimation of the optical intensity relevant to each component. The spectrum of the light collected by the telescope can be derived from the data coming out from the three spectrometers.

The key element of each spectrometer is the grating, which splits and diffracts the incident light into several beams propagating in different directions. Since the directions depend on the wavelength, the grating acts as the dispersive element enabling the spectrometer operation. Figure 6.15 schematically shows the functionality of the grating.

The spectrometers included in ChemCam are Czerny–Turner spectrometers [84] (see Fig. 6.16). The incoming light is collimated by a concave mirror and directed towards the grating; from the grating the

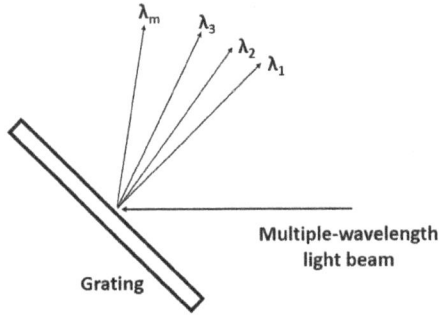

Fig. 6.15. Scheme showing the principle of a grating acting as a dispersive element.

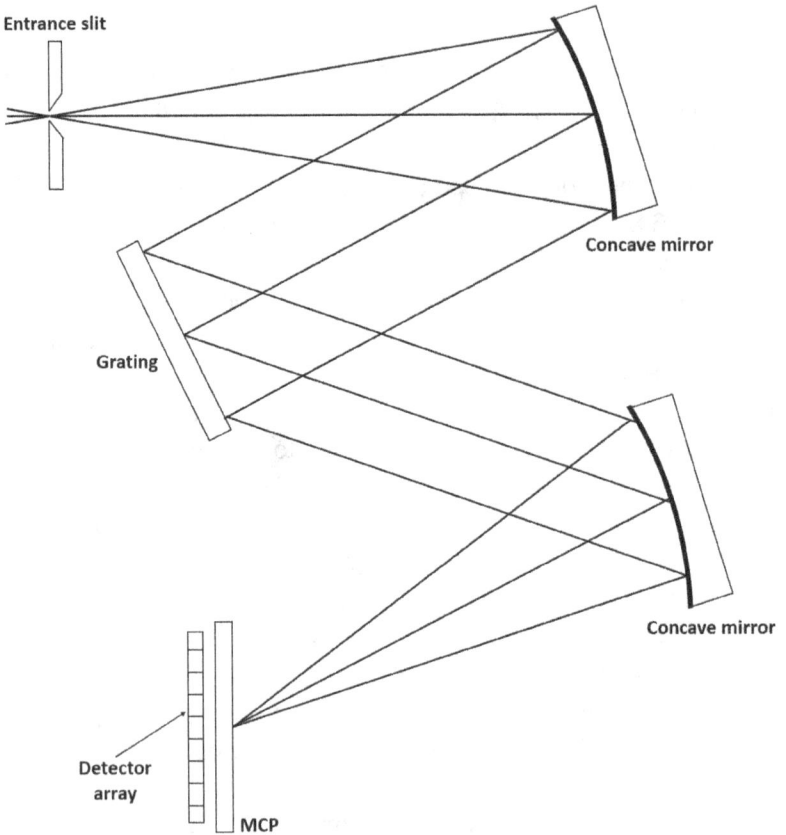

Fig. 6.16. Configuration of the Czerny–Turner included in ChemCam. MPC: micro-channel plate.

dispersed radiation reaches the CCD-based detector array though another concave mirror. The spectrometer resolution is 0.2 nm below 500 nm and 0.65 nm above 500 nm.

LIBS provides information on the elemental composition of the Mars surface and rocks but it is not able to detect low levels (concentration <1 ppm) of organics [85]. This functionality can be accomplished by Raman spectroscopy, which implies the spectral analysis of the light scattered by the target, when a laser beam is directed towards it. In particular, for this kind of spectroscopy, the scattered light of interest is the one which is frequency shifted from the laser frequency due to interactions between the incident beam and the vibrational energy levels of the molecules in the sample. A Raman spectrometer, whose TRL is 5, is currently under development to be included in the planetary rover that will explore Mars in the framework of the ESA mission ExoMars [86]. The excitation laser is at 532 nm and it generates a beam that allows an irradiance on the target between 0.8 kW/cm^2 and 1.2 kW/cm^2.

6.4 Remote sensing instruments

Observation of the Earth or the other planets of the Solar Systems is the goal of a lot of space missions. Observation is performed by remote sensing instruments, which often use light to obtain information on the specific mission target. Optical remote sensing can be either passive, when no self-generated radiation is used in the sensing, or active, if the instrument generates a laser beam, which is then directed towards the target.

In the last few decades, several optical remote sensing instruments, e.g. interferometers and spectrometers, have been developed in the framework of many space missions. For example, one of the two orbiters that ESA will launch towards Mercury, the Mercury Planetary Orbiter (MPO), in the framework of the BepiColombo mission (planned launch: January 2017), will include an optical instrument, called PHEBUS (Probing of Hermean Exosphere by Ultraviolet Spectroscopy). It will be utilized to study the composition and the dynamics of the Mercury exosphere through extreme ultraviolet (EUV) and the far ultraviolet (FUV) spectroscopy. PHEBUS (mass about 7 kg), which performs passive remote sensing, consists of two spectrophotometers operating in two different wavelength ranges, from 55 nm to 155 nm (EUV) and from 145 nm to 315 nm (FUV), and a scanning mirror, collecting the light coming from the Mercury exosphere. The key elements of the spectrometers are the gratings, acting as dispersive elements, and the detectors, which are both based on microchannel plate technology. The instrument also includes two small near-UV detectors for the quantification of the light from calcium and potassium molecules.

The most common optical technique for active remote sensing from Space is Light Detection and Ranging (LIDAR) [87]. The LIDAR instrument sends a laser beam towards the target and collects/analyses the reflected and/or backscattered radiation using a telescope and a detector. It can be designed either to measure the distance of a hard target or to characterize the back-scattering due to soft targets, such as aerosols and clouds. In the former case, the instrument, which is usually called a laser altimeter, allows the reconstruction of the global shape and the surface elevation of a celestial body, while in the latter case, the LIDAR technique is mainly used to study the Earth's atmosphere.

A block diagram of a LIDAR system is shown in Fig. 6.17. The system includes three fundamental building blocks, i.e. a transmitter, a receiver, and a detector. The transmitter consists of a laser source, emitting optical pulses in the visible or in the near-infrared, and a beam expander, which reduces the divergence of the transmitted light. In the receiver, the light is collected by a telescope and then it is processed in the optical domain. In the detector, the received light pulses are transduced in the electrical domain and then processed by the instrument electronics.

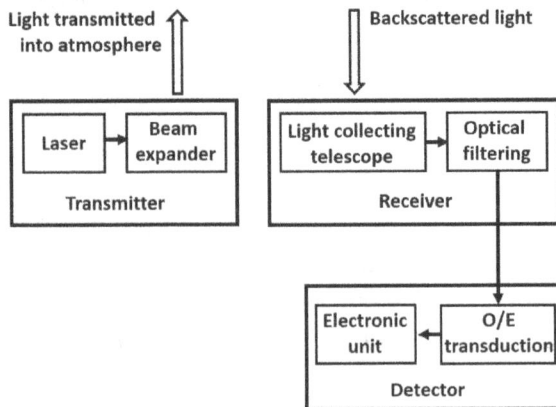

Fig. 6.17.   Block diagram of a LIDAR system.

Laser altimeters [88], which were used in Space for the first time in the Apollo 15 mission (launched on July 26, 1971), sent a train of short pulses towards the target and measured the time-of-flight, i.e. the time interval between the pulse launch and the reception of the reflected replica of the same pulse. In addition, the altimeters often monitored both the echo pulse energy and the pulse spreading caused by the surface of a celestial body under investigation, e.g. Mars or the Moon.

Detailed topographic maps of Mars, Earth, the Moon, Mercury (northern hemisphere), and a pair of asteroids (433 Eros and Itokawa) are currently available [89] thanks to the measurements performed by spaceborne LIDAR instruments, which typically use Q-switched Nd:YAG lasers pumped by laser diode arrays and silicon avalanche photodiodes as detectors. The transmitted pulses have a temporal width of some nanoseconds while the repetition rate is of the order of 10 Hz.

One of the most advanced laser altimeters among those currently operating is the lunar orbiter laser altimeter (LOLA) [90], shown in Fig. 6.18. The instrument, which was launched in June 2009 on board of the Lunar Reconnaissance Orbiter, is the first multi-beam laser altimeter

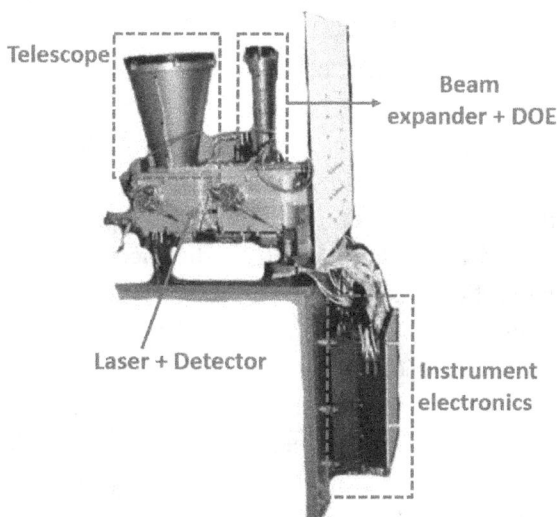

Fig. 6.18.   LOLA instrument.
Source: Adapted with permission from Fig. 17 of Ref. [90].

operating in space. The altimeter is aimed at providing a geodetic map of lunar topography with a resolution of 25 m.

The laser transmitter [91], emitting pulses having an energy of about 3 mJ and a temporal width of 6 ns, with repetition rate of 28 Hz, is based on a Q-switched diode pumped solid-state laser emitting at 1,064 nm. The peculiar feature of this instrument is the use of a diffractive optical element (DOE) [92], which splits the laser-generated beam into five separate beams at the output of the beam expander. The element is a grating, which modifies the phase of the incoming light to produce the desired far field pattern, manufactured by selectively etching the surface of a fused silica substrate. Due to this diffractive optical element, each optical pulse produces five laser spots on the lunar surface, the height, slope, and orientation of which can be simultaneously evaluated.

The reflected light is collected by a telescope with a diameter of 14 cm [93]. Five optical fibers are located at the telescope focal plane, each aligned to one of the five laser spots on the lunar surface. In the fibers, the echo optical pulses propagate towards the five avalanche photodiodes.

The laser altimeters that are currently operating do not utilize the digital detection of the echo pulse, the leading-edge of which is detected by analog techniques. This digital functionality is implemented in the laser altimeter BELA, BepiColombo Laser Altimeter [93], which will be mounted on board the MPO, one of the two orbiter that the ESA will launch towards Mercury.

BELA is the first European laser altimeter for planetary observation and it includes a Q-switched Nd:YAG laser at 1,064 nm, generating optical pulses with an energy of up to 50 mJ and a repetition rate of 10 Hz, which allow the instrument to operate at distances exceeding 1,000 km from the target surface. As in all spaceborne altimeters, detection is performed by silicon-based avalanche photodiodes.

An atmospheric aerosol [94] is a mixture of solid and liquid particles, with diameter from a few nm to a few hundred μm, which are suspended in the atmosphere. It originates from natural and anthropogenic sources and exhibits, together with clouds, a strong influence on the Earth's climate.

Spaceborne LIDAR instruments are emerging as powerful tools for characterizing both aerosol and clouds. In fact, the Cloud-Aerosol Lidar with Orthogonal Polarization (CALIOP) instrument [95], which is mounted on board the Cloud-Aerosol Lidar and Infrared Pathfinder Satellite Observations (CALIPSO) satellite (launch: April 28, 2006), has demonstrated its ability to provide high resolution vertical profiles of clouds and aerosol.

The transmitter of CALIOP [96], which is based on a diode-pumped Q-switched Nd:YAG laser at 1,064 nm, generates simultaneous and co-aligned optical pulses at 1,064 nm and 532 nm, each having an energy of about 100 mJ. The pulse repetition rate is about 20 Hz and the pulse duration is 20 ns. The pulses at 532 nm are generated by a frequency doubling crystal. The light backscattered by clouds and aerosol is collected by a telescope having a diameter of 1 m. At the telescope output, a dichroic beam splitter divides the light at 532 nm from that at 1,064 nm. Both the so obtained beams are filtered and the backscattered radiation at 532 nm pass through a polarization beam splitter that separates the light polarized perpendicular to the polarization direction of the transmitted beam from the light polarized parallel to the polarization direction of the transmitted beam. The intensity of the two beams at 532 nm is detected by two photomultipliers while a silicon avalanche photodiode detects the beam at 1064 nm. A block diagram of CALIOP, mass = 172 kg and power consumption = 197 W, is shown in Fig. 6.19.

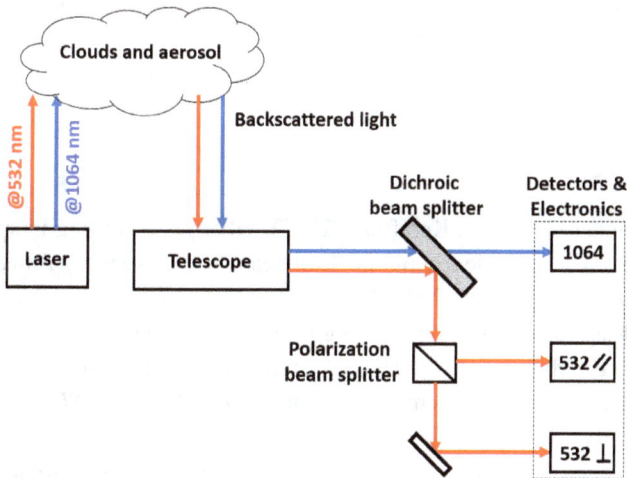

Fig. 6.19.   Block diagram of CALIOP.

The estimation of the polarization conversion due to the interaction of the pulses at 532 nm with aerosol/clouds allows the discrimination between spherical and non-spherical particles while the use of two different wavelengths provides information on aerosol/cloud particle size [97].

In the near future, ESA has planned to launch two important missions to study both the vertical wind profiles in the troposphere and lower stratosphere and the aerosol–cloud-radiation interactions. The missions, called Atmospheric Dynamics Mission-Aeolus [98] and EarthCARE, Earth Clouds Aerosols and Radiation Explorer [99], are both based on LIDAR instruments.

The LIDAR technique can be used to estimate the concentration of some gases, such as carbon dioxide, ozone, and methane, within the Earth's atmosphere. Those measurements can be performed using differential absorption LIDAR (DIAL) [100], which utilizes a pulsed laser emitting at two different wavelengths. The DIAL instrument evaluates the amount of light backscattered by the atmosphere at two specific wavelengths, one (on-wavelength) relevant to an absorption line of the measured chemical species (e.g. 1645.552 nm for methane sensing) and the other one (off-wavelength) located outside the absorption lines. Although several space agencies have envisaged DIAL missions, e.g. the ESA A-SCOPE [101], Advanced Space Carbon and Climate Observation of Planet Earth, the Franco-German MERLIN [102], NASA ASCENDS [103], and, Active Sensing of CO_2 Emissions over Nights, Days, and Seasons, to date, DIAL instruments have never been operated in Space.

References

[1] I. Mckenzie and N. Karafolas (2005). Fiber optic sensing in space structures: The experience of the European Space Agency, *Proceedings of SPIE*, vol. 5855, pp. 262–269.

[2] R. Kashyap (2010). *Fiber Bragg Gratings*. Elsevier, Amsterdam.

[3] P. J. Ellerbrock (1997). DC-XA structural health monitoring fiber-optic based strain measurement system, *Proceedings of SPIE*, vol. 3044, pp. 207–218.

[4] J. Sirkis, B. Childers, L. Melvin, T. Peng, Yi Tang, John J. Moore, E. Enright and C. Bovier (1999). Integrated vehicle health management

(IVHM) on space vehicles: A Space shuttle flight experiment, *Key Engineering Materials*, vol. 167–168, pp. 273–280.

[5] N. Takeda, T. Mizutani, K. Hayashi and Y. Okabe (2003). Application of fiber Bragg grating sensors to real-time strain measurement of cryogenic tanks, *Proceedings of SPIE*, vol. 5056, p. 304.

[6] A. Reutlinger, M. Glier, K.-H. Zuknik, L. Hoffmann, M. Müller, S. Rapp, C. Kurvin, T. Ernst, I. McKenzie and N. Karafolas (2008). Fiber optic sensing for telecommunication satellites, *Proceedings of SPIE*, vol. 7004, p. 700454.

[7] S. Abad, F. M. Araújo, L. A. Ferreira, F. Pedersen, M. A. Esteban, I. Mckenzie and N. Karafolas (2010). Applications of FBG Sensors on Telecom Satellites, *International Conference on Space Optics*, Rhodes, Greece, October 4–8.

[8] K. Gantois, F. Teston, O. Montenbruck, P. Vuilleumier, P. V. D. Braembussche and M. Markgraf (2006). PROBA-2 mission and new technologies overview, *4S Symposium on Small Satellite Systems and Services*, Chia Laguna, Sardinia, Italy, September 25–29.
   L. Maleki (2011). The optoelectronic oscillator, *Nature Photonics*, vol. 5, pp. 728–730.

[9] R. V. Kruzelecky, J. Zou, N. Mohammed, E. Haddad, W. Jamroz, F. Ricci, J. Lamorie, E. Edwards, I. McKenzie and P. Vuilleumier (2006). Fiber-Optic Sensor Demonstrator (FSD) for the Monitoring of Spacecraft Subsystems on ESA's Proba-2, *International Conference on Space Optics*, ESTEC, Noordwijk, The Netherlands, June 27–30.

[10] K. O. Hill and G. Meltz (1997). Fiber Bragg Grating technology fundamentals and overview, *Journal of Lightwave Technology*, vol. 15, pp. 1263–1276.

[11] P. Putzer, A. W. Koch, M. Plattner, A. Hurni and M. Manhart (2010). Design of a Fiber-Optic Interrogator Module for Telecommunication Satellites, *International Conference on Space Optics*, Rhodes, Greece, October 4–8.

[12] A. Kersey, M. A. Davis, H. J. Patrick, M. Leblanc, K. P. Koo, C. G. Askins, M. A. Putnam and E. J. Friebele. Fiber grating sensors, *Journal of Lightwave Technology*, vol. 15, pp. 1442–1463.

[13] H. Czichos, T. Saito and L. Smith (2006). *Springer Handbook of Materials Measurement Methods*, Chapter 11. Springer, New York.

[14] S. J. Mihailov, C. W. Smelser, P. Lu, R. B. Walker, D. Grobnic, H. Ding, G. Henderson and J. Unruh (2003). Fiber Bragg gratings made with a

phase mask and 800-nm femtosecond radiation, *Optics Letters*, vol. 28, pp. 995–997.

[15] A. Martinez, M. Dubov, I. Khrushchev and I. Bennion (2004). Direct writing of fibre Bragg gratings by femtosecond laser, *Electronics Letters*, vol. 40, pp. 1170–1172.

[16] A. I. Gusarov, F. Berghmans, A. F. Fernandez, O. Deparis, Y. Defosse, D. Starodubov, M. Decreton, P. Megret and M. Bondel. Behavior of fibre Bragg gratings under high total dose gamma radiation, *IEEE Transactions on Nuclear Science*, vol. 47, pp. 688–692.

[17] D. Grobnic, H. Henschel, S. K. Hoeffgen, J. Kuhnhenn, S. J. Mihailov and U. Weinand (2009). Radiation sensitivity of Bragg gratings written with femtosecond IR lasers, *Proceedings of SPIE*, vol. 7316, 73160.

[18] H. Henschel, D. Grobnic, S. K. Hoeffgen, J. Kuhnhenn, S. J. Mihailov and U. Weinand (2011). Development of highly radiation resistant fiber Bragg gratings, *IEEE Transactions on Nuclear Science*, vol. 58, pp. 2103–2110.

[19] M. Perry, P. Niewczas and M. Johnston (2012). Effects of neutron-gamma radiation on fiber bragg grating sensors: A review, *IEEE Sensors Journal*, vol. 12, pp. 3248–3257.

[20] M. A. Aguirre (2012). *Introduction to Space Systems: Design and Synthesis*. Springer, New York.

[21] B. N. Agrawal and W. J. Palermo (2002). Angular Rate Estimation for Gyroless Satellite Attitude Control, AIAA Guidance, Navigation, and Control Conference, Monterey, California, USA, August, 5–8.

[22] M. N. Armenise, C. Ciminelli, F. Dell'Olio and V. M. N. Passaro (2011). *Advances in gyroscope technologies*. Springer, New York.

[23] K. S. Ali, C. A. Vanelli, J. J. Biesiadecki, M. W. Maimone, Y. Cheng, A. M. San Martin and J. W. Alexander (2005). Attitude and Position Estimation on the Mars Exploration Rovers, *IEEE International Conference on Systems, Man and Cybernetics*, The Big Island, HI, USA, October 10–12.

[24] S. Kowaltschek (2012). Lessons learnt from the SiREUS MEMS detector evaluation, *6th ESA Workshop on Avionics Data, Control and Software Systems*, ESA/ESTEC, Noordwijk, The Netherlands, October 23–25.

[25] W. M. Macek and D.T.M. Davis (1963). Rotation rate sensing with travelling-wave ring lasers, *Applied Physics Letters*, vol. 2, pp. 67–68.

[26] D. H. Titterton and J. L. Weston (2004). *Strapdown Inertial Navigation Technology*, The Institution of Electrical Engineers.

[27] M. Faucheux, D. Fayoux and J. J. Roland (1988). The ring laser gyro, *Journal of Optics*, vol. 19, pp. 101–115.

[28] S. Schwartz, G. Feugnet, P. Bouyer, E. Lariontsev, A. Aspect and J.-P. Pocholle (2006). Mode coupling control in resonant devices: Application to solid-state ring lasers, *Physics Review Letters*, vol. 97, pp. 093902.

[29] F. Aronowitz (1999). Fundamentals of the ring laser gyro, in *Optical Gyros and their Applications*, D. Loukianov, R. Rodloff, H. Sorg and B. Stieler (Eds.), NATO Research and Technology Organization.

[30] J. Killpatrick (1969). Random bias for laser angular rate sensor, US Patent 3,467,472.

[31] W. W. Chow, J. B. Hambenne, T. J. Hutchings, V. E. Sanders, M. Sargent and M. O. Scully (1980). Multioscillator laser gyros, IEEE *Journal of Quantum Electronics*, vol. QE-16, pp. 918–936.

[32] J. J. Roland and G. P. Agrawal (1981). Optical gyroscopes, *Optics & Laser Technology*, vol. 13, pp. 239–244.

[33] R. L. Brown (1968). NRL Memorandum Report N1871, Naval Research Laboratory, Washington.

[34] R. Ulrich (1980). Fiber-optic rotation sensing with low drift, *Optics Letters*, vol. 5, pp. 173–175.

[35] C. Ciminelli, F. Dell'Olio, C. E. Campanella and M. N. Armenise (2010). Photonic technologies for angular velocity sensing, *Advances in Optics and Photonics*, vol. 2, pp. 370–404.

[36] N. J. Frigo (1983). Compensation of Linear Sources of Non-Reciprocity in Sagnac Interferometers, *Proceedings of SPIE*, vol. 0412, pp. 268–271.

[37] Datasheet of the RGA-14 RLG, L-3 Communications, http://www.l-3com. com/.

[38] Datasheet of the RGA-20 RLG, L-3 Communications, http://www.l-3com. com/.

[39] Airbus Defence & Space, GNC Sensors & Actuators, http://www.space-airbusds.com/en/equipment/about-sensors.html.

[40] Datasheet of the Miniature inertial measurement unit (MIMU), Honeywell, http://honeywell.com/.

[41] Datasheet of the LN-200 FOG Family, Northrop Grumman, http://www. northropgrumman.com/.

[42] R. E. Meyer, S. Ezekiel, D. W. Stowe and V. J. Tekippe (1983). Passive fiber-optic ring resonator for rotation sensing, *Optics Letters*, vol. 8, pp. 644–646.

[43] G. A. Sanders, N. Demma, G. F. Rouse and R. B. Smith (1988). Evaluation of Polarization Maintaining Fiber Resonator for Rotation Sensing Applications, Optical Fiber Sensors, New Orleans, LA, USA, January 27.

[44] X. Zhang, H. Ma, Z. Jin and C. Ding (2006). Open-loop operation experiments in a resonator fiber-optic gyro using the phase modulation spectroscopy technique, *Applied Optics*, vol. 45, pp. 7961–7965.

[45] K. Iwatsuki, K. Hotate and M. Higashiguchi (1984). Effect of Rayleigh backscattering in an optical passive ring-resonator gyro, *Applied Optics*, vol. 23, pp. 3916–3924.

[46] K. Iwatsuki, K. Hotate and M. Higashiguchi (1986a). Eigenstate of polarization in a fiber ring resonator and its effect in an optical passive ring-resonator gyro, *Applied Optics*, vol. 25, pp. 2606–2612.

[47] H. Ma, Z. Chen, Z. Yang, X. Yu and Z. Jin (2012). Polarization-induced noise in resonator fiber optic gyro, *Applied Optics*, vol. 51, pp. 6708–6717.

[48] K. Iwatsuki, H. Hotate and M. Higashiguchi (1986b). Kerr effect in an optical passive ring-resonator gyro, *Journal of Lightwave Technology*, vol. 4, pp. 645–651.

[49] M. A. Terrel, M. J. F. Digonnet and S. Fan (2012). Resonant fiber optic gyroscope using an air-core fiber, Journal of Lightwave Technology, vol. 30, pp. 931–937.

[50] T. Imai, Y. Miki, S. Maeda and K. Nishide (1996). Development of resonator fiber optic gyros, Optical Fiber Sensors, Sapporo, Japan, May 21–24.

[51] C. Ciminelli, F. Dell'Olio, M. N. Armenise (2013). Resonant optical gyro: Monolithic *vs.* hybrid integration, *International Conference on Transparent Optical Networks (ICTON 2013)*, Cartagena, Spain, June 23–27.

[52] M. Armenise and P. J. R. Laybourn (1998). Design, simulation of a ring laser for miniaturized gyroscopes, *Proceedings of SPIE*, vol. 3464, pp. 81–90.

[53] M. N. Armenise, M. Armenise, V. M. N. Passaro and F. De Leonardis (2000). Integrated optical angular velocity sensor, European Patent 1219926.

[54] C. Ciminelli, F. Dell'Olio, C. E. Campanella, V. M. N. Passaro and M. N. Armenise (2010). Integrated Optical Ring Resonators: Modelling and Technologies, in *Progress in Optical Fibers*, P. S. Emersone (Ed.), Nova Science Publisher, New York.

[55] F. Dell'Olio, T. Tatoli, C. Ciminelli and M. N. Armenise (2014). Recent advances in miniaturized optical gyroscopes, *Journal of the European Optical Society — Rapid Publications*, vol. 9, p. 14013.

[56] M. N. Armenise, C. Ciminelli, F. Dell'Olio, A. Di Nisio, M. Savino and M. Spadavecchia (2015). Out-of-Resonance Measurement Scheme for Ring Resonator Gyroscopes, *IEEE International Instrumentation and Measurement Technology Conference (I2MTC)*, Pisa, Italy, May 11–14.

[57] J. Wang, L. Feng, Y. Tang and Y. Zhi (2015). Resonator integrated optic gyro employing trapezoidal phase modulation technique, *Optics Letters*, vol. 40, pp. 155–158.

[58] Y. Zhi, L. Feng, J. Wang and Y. Tang (2015). Reduction of backscattering noise in a resonator integrated optic gyro by double triangular phase modulation, *Applied Optics*, vol. 54, pp. 114–122.

[59] S. Ezekiel and S. R. Balsamo (1977). Passive ring resonator laser gyroscope, *Applied Physics Letters*, vol. 30, pp. 478–480.

[60] C. Ciminelli, F. Dell'Olio, C. E. Campanella and M. N. Armenise (2012a). High-Q spiral resonator for optical gyroscope applications: Numerical and experimental investigation, *IEEE Photonics Journal*, vol. 4, pp. 1844–1854.

[61] C. Ciminelli, F. Dell'Olio, C. E. Campanella and M. N. Armenise (2012b). Numerical and Experimental Investigation of an Optical High-Q Spiral Resonator Gyroscope, *International Conference on Transparent Optical Networks (ICTON)*, University of Warwick, Coventry, UK, July 2–5.

[62] H. Mao, H. Ma and Z. Jin (2011). Polarization maintaining silica wave-guide resonator optic gyro using double phase modulation technique, *Optics Express*, vol. 19, pp. 4632–4643.

[63] C. Ciminelli, C. E. Campanella and M. N. Armenise (2014). Optical rotation sensor as well as method of manufacturing an optical rotation sensor, European Patent 056933.

[64] F. Dell'Olio, C. Ciminelli and M. N. Armenise (2013). Theoretical investigation of InP buried ring resonators for new angular velocity sensors, *Optical Engineering*, vol. 52, p. 024601.

[65] C. Ciminelli, V. M. N. Passaro, F. Dell'Olio and M. N. Armenise (2009). Quality factor and finesse optimization in buried InGaAsP/InP ring resonators, *Journal of the European Optical Society — Rapid Publications*, vol. 4, p. 09032.

[66] F. Dell'Olio, C. Ciminelli, M. N. Armenise, F. M. Soares and W. Rehbein (2012). Design, Fabrication, and Preliminary Test Results of a New InGaAsP/InP High-Q ring Resonator for Gyro Applications, 24th *International Conference on Indium Phosphide and Related Materials (IPRM)*, Santa Barbara, CA, USA, August 27–30.

[67] C. Ciminelli, F. Dell'Olio, M. N. Armenise, F. M. Soares and W. Passenberg (2013). High performance InP ring resonator for new generation monolithically integrated optical gyroscopes, *Optics Express*, vol. 21, pp. 556–564.

[68] F. Dell'Olio, F. Indiveri, C. Ciminelli and M. N. Armenise (2014). Optoelectronic gyroscope based on a high-Q InGaAsP/InP ring resonator: Preliminary results of the system test, *16th International Conference on Transparent Optical Networks (ICTON)*, Graz, Austria, July 6–10.

[69] F. Dell'Olio, A. Di Nisio, F. Indiveri, P. Lino, C. Ciminelli and M. N. Armenise (2014). Backscattering noise control in the readout circuit of innovative optoelectronic resonant gyroscopes, *AEIT Italian Conference on Photonics Technologies*, Fotonica, Naples, Italy, May 12–14.

[70] F. Dell'Olio, F. Indiveri, F. Innone, C. Ciminelli and M. N. Armenise (2014). Experimental countermeasures to reduce the backscattering noise in an InP hybrid optical gyroscope, *Third Mediterranean Photonics Conference*, Trani, Italy, May 7–9.

[71] F. Dell'Olio, F. Indiveri, F. Innone, P. Dello Russo, C. Ciminelli and M. N. Armenise (2014). System test of an optoelectronic gyroscope based on a high Q-factor InP ring resonator, *Optical Engineering*, vol. 53, 127104.

[72] J. Wang, L. Feng, Y. Tang and Y. Zhi (2015). Resonator integrated optic gyro employing trapezoidal phase modulation technique, *Optics Letters*, vol. 40, pp. 155–158.

[73] B. Z. Steinberg, J. Scheuer and A. Boag (2007). Rotation-induced super-structure in slow-light waveguides with modedegeneracy: Optical gyroscopes with exponential sensitivity, *Journal of the Optical Society of America B*, vol. 24, pp. 1216–1224.

[74] M. A. Terrel, M. J. F. Digonnet and S. Fan (2009). Performance limitation of a coupled resonant optical waveguide gyroscope, *Journal of Lightwave Technology*, vol. 27, pp. 47–54.

[75] J. R. E. Toland, Z. A. Kaston, C. Sorrentino and C. P. Search (2011). Chirped area coupled resonator optical waveguide gyroscope, *Optics Letters*, vol. 36, pp. 1221–1223.

[76] D. Kalantarov and C. P. Search (2013). Effect of input–output coupling on the sensitivity of coupled resonator optical waveguide gyroscopes, *Journal of the Optical Society of America B*, vol. 30, pp. 377–381.

[77] C. Sorrentino, J. R. E. Toland and C. P. Search (2012). Ultra-sensitive chip scale Sagnac gyroscope based on periodically modulated coupling of a coupled resonator optical waveguide, *Optics Express*, vol. 20, pp. 354–363.

[78] P. S. Schenker, T. L. Huntsberger, P. Pirjanian, E. T. Baumgartner and E. Tunstel (2003). Planetary rover developments supporting mars exploration, sample return and future human-robotic colonization, *Autonomous Robots*, vol. 14, pp. 103–126.

[79] S. Stenholm (2005). *Foundations of Laser Spectroscopy*. Dover Publications, New York.

[80] S. Maurice *et al.* (2012). The ChemCam instrument suite on the Mars Science Laboratory (MSL) Rover: Science objectives and mast unit description, *Space Science Reviews*, vol. 170, pp. 95–166.

[81] R. Welch, D. Limonadi and R. Manning (2013). Systems engineering the Curiosity Rover: A retrospective, *8th International Conference on System of Systems Engineering (SoSE)*, Maui, HI, USA, June 2–6.

[82] R. Noll (2012). *Laser-Induced Breakdown Spectroscopy: Fundamentals and Applications*. Springer, New York.

[83] B. Faure, E. Durand, S. Maurice, D. Bruneau and F. Montmessin (2012). New Developments on ChemCam Laser Transmitter and Potential Applications for other Planetology Programs, *International Conference on Space Optics*, Ajaccio, Corse, October 9–12.

[84] M. Czerny and A. F. Turner (1930). On the astigmatism of mirror spectrometers, *Z. Phys.*, vol. 61, pp. 792–797.

[85] S. M. Angel, N. R. Gomer, S. K. Sharma and C. McKay (2012). Remote Raman spectroscopy for planetary exploration: A review, *Applied Spectroscopy*, vol. 66, pp. 137–150.

[86] F. Rull, S. Maurice, E. Diaz, C. Tato and A. Pacros (2011). The Raman Laser Spectrometer (RLS) on the ExoMars 2018 Rover Mission, *42nd Lunar and Planetary Science Conference*, The Woodlands, Texas, USA, March 7–11.

[87] C. Weitkamp (2005). Lidar: Introduction, in *Laser Remote Sensing*, T. Fujii and T. Fukuchi (Eds.), Taylor & Francis, Oxford.

[88] J. B. Abshire (2011). NASA's Space Lidar Measurements of the Earth and Planets, *IEEE Photonics Society Meeting*, University of Maryland, April 5.

[89] X. Sun (2012). Space-Based Lidar Systems, *Conference on Lasers and Electro-Optics*, Paper JW3C.5, Baltimore, Maryland, USA, May 6–11.

[90] X. Sun, J. B. Abshire, J. F. McGarry, G. A. Neumann, J. C. Smith, J. F. Cavanaugh, D. J. Harding, H. J. Zwally, D. E. Smith and M. T. Zuber (2013). Space Lidar Developed at the NASA Goddard Space Flight Center — The First 20 Years, *IEEE Journal of Selected Topics in Applied Earth Observations and Remote Sensing*, vol. 6, pp. 1660–1675.

[91] A. W. Yu, A. M. Novo-Gradac, G. B. Shaw, S. X. Li, D. C. Krebs, L. A. Ramos-Izquierdo, G. L. Unger and A. Lukemire (2008). Laser Transmitter for the Lunar Orbit Laser Altimeter (LOLA) Instrument, *Conference on Lasers and Electro-Optics/Quantum Electronics (CLEO)*, San Jose, CA, USA, May 4–9.

[92] A. W. Yu, A. M. Novo-Gradac, G. B. Shaw, S. X. Li, D. C. Krebs, L. A. Ramos-Izquierdo, G. L. Unger and A. Lukemire (2009). Laser Transmitter for the Lunar Orbit Laser Altimeter (LOLA) Instrument, *Conference on Lasers and Electro-Optics/Quantum Electronics (CLEO)*, Baltimore, MD, USA, June 2–4.

[93] N. Thomas *et al.* (2007). The BepiColombo Laser Altimeter (BELA): Concept and baseline design, *Planetary and Space Science*, vol. 55, pp. 1398–1413.

[94] M. Chin *et al.* (2009). *Atmospheric Aerosol Properties and Climate Impacts*. DIANE Publishing, Darby.

[95] D. M. Winker, W. H. Hunt and M. J. McGill (2007). Initial performance assessment of CALIOP, *Geophysical Research Letters*, vol. 34, p. L19803.

[96] D. M. Winker, J. Pelon and M. P. McCormick (2003). The CALIPSO mission: Spaceborne LIDAR for observation of aerosols and clouds, *Proceedings of SPIE*, vol. 4893, pp. 1–11.

[97] Y.-S. Choi, R. S. Lindzen, C.-H. Ho and J. Kim (2010). Space observations of cold-cloud phase change, *PANAS*, vol. 107, pp. 11211–11216.

[98] ESA Earth Online, ADM-Aeolus, https://earth.esa.int/web/guest/missions/esa-future-missions/adm-aeolus.

[99] ESA Earthcore, http://www.esa.int/Our_Activities/Observing_the_Earth/The_Living_Planet_Programme/Earth_Explorers/EarthCARE/ESA_s_cloud_aerosol_and_radiation_mission.

[100] U. Platt and J. Stutz (2008). *Differential Optical Absorption Spectroscopy: Principles and Applications*. Springer, New York.

[101] Y. Durand, J. Caron, P. Bensi, P. Ingmann, J.-L. Bézy and R. Meynart (2009). A-SCOPE: Objectives and Concepts for an ESA Mission to Measure CO_2 from Space with a LIDAR, *15th Coherent Laser Radar Conference*, CLRC XV, Toulouse, France, June 22–26, 2009.

[102] eoPortal Directory, Merlin, https://directory.eoportal.org/web/eoportal/satellite-missions/m/merlin.

[103] J. B. Abshire *et al.* (2010). Pulsed airborne LIDAR measurements of atmospheric CO_2 column absorption, *Tellus B*, vol. 62, pp. 770–783.

Chapter 7

Solar Cells for Space

For some decades, solar cells have played an important role in providing power to the International Space Station, satellites and planetary rovers. Since 1958, when the first satellite powered by solar cells, Vanguard 1, was launched, remarkable developments in the solar cells for Space have taken place, with efficiency increasing from 10% in the first cells to about 30% in the current ones [1].

Solar cells convert the electromagnetic radiation emitted by the Sun into electric power, by exploiting the photovoltaic effect. The two physical processes involved in the conversion of sunlight into electrical energy are the absorption of the Sun-generated electromagnetic radiation within a light absorbing semiconductor material, with the generation of electron–hole pairs, and the separation of electrons and holes due to the electric field across a semiconductor p–n junction, originating a photocurrent.

Since a solar cell is basically a p–n junction under solar irradiation, its I–V characteristic is [2]:

$$I = I_s \exp\left(\frac{qV}{kT} - 1\right) - I_L,$$

(7.1)

where I_L is the current due to the Sun-generated photons, I_S is the junction saturation current, k is the Boltzmann's constant, and T is the absolute temperature.

199

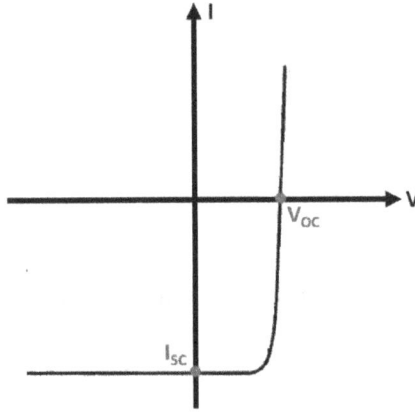

Fig. 7.1.   *I–V* characteristic of a *p–n* junction under solar irradiation.

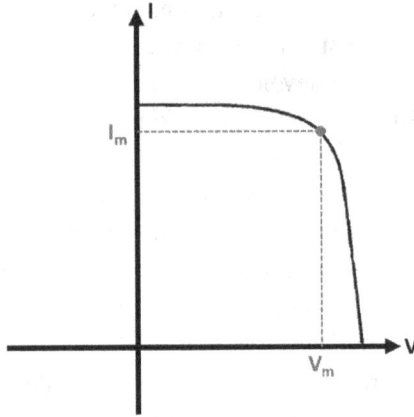

Fig. 7.2.   Typical *I–V* characteristic of a solar cell.

Figure 7.1 shows a qualitative plot of the *I–V* characteristic, which passes through the fourth quadrant. This means that the junction is able to generate electrical power. The current through the junction when the voltage across it is zero is the short-circuit current I_{SC}, while the voltage for $I = 0$ is the open-circuit voltage V_{OC}. The optimum value of the load, defined as the one maximizing the power generated by the cell, is about 80% of the product $I_{SC}V_{OC}$.

The *I–V* characteristic of a solar cell in Fig. 7.2 is obtained from Fig. 7.1 by an inversion of the plot about the voltage axis. The voltage V_m and the

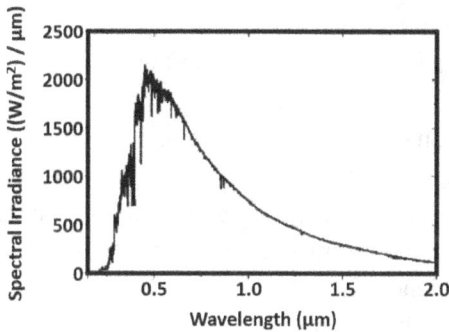

Fig. 7.3.   Solar irradiance spectrum outside the atmosphere (AMO).

current I_m are defined as the voltage and current values allowing the maximization of the output power. The maximum generated power is denoted as $P_m = V_m I_m$.

The solar spectral irradiance (power per unit area, per unit wavelength) outside the Earth's atmosphere, which is denoted as the AMO spectrum [3], is shown in Fig. 7.3. That spectrum, which is used to evaluate the performance of solar cells intended for space, shows that the energy in sunlight is distributed over a broad spectrum of wavelengths. This is the reason why the first Si-based solar cells for space, which were only able to convert into electrical energy the radiation with wavelength < 1.1 μm, are not utilized by the space industry because of their poor efficiency. Currently, only the multi-junction solar cells based on III–V materials, having a record efficiency >35% and a very good resistance to radiation, are utilized in spacecraft engineering.

In this chapter, we will discuss the technology of III–V multi-junction solar cells for space applications, with some remarks on solar arrays and the future trends of this technology.

7.1 III–V multi-junction solar cells for space

In solar cells based on a single p–n junction in Si or GaAs, the Sun-generated photons having energy less than the material band gap energy E_g give no contribution to the cell output, while for photons having energy $h\nu > E_g$, the excess energy $h\nu - E_g$ is lost as heat. This obviously limits the maximum achievable efficiency of these devices.

In solar cells including vertically stacked junctions with different band gaps, the efficiency can be increased with respect to the single junction cells because there is less waste of photons.

In multi-junction solar cells, the different junctions are arranged in a stack where the band gaps decrease downward into the stack. A Sun-generated photon entering the stack from the top passes through the stack until it reaches the junction that is able to convert it into an electron–hole couple.

The high efficiency demonstrated by III–V multi-junction solar cells is mainly due to the band gap tunability of the III–V material system [4]. In fact, by varying the stoichiometry of the III–V compounds, the band gap and the lattice constant of these materials can be easily engineered. Figure 7.4 shows the band gap energy versus lattice constant for most III–V compound semiconductors, Ge and Si, and illustrates the wide range of tunability of both band gap and lattice constant in III–V materials.

The key criteria in the selection of the materials for the multi-junction solar cells are the optimization of the band gap values, taking into account the spectrum of the Sun irradiance, and the minimization of crystallographic defects, i.e. impurities, dislocations, and fractures, within the multi-layer structure. The band gap optimization is aimed at the generation of the same photocurrent in each junction while the reduction of the defect density can be achieved by minimizing the number of recombination sites for photo generated electron–hole pairs.

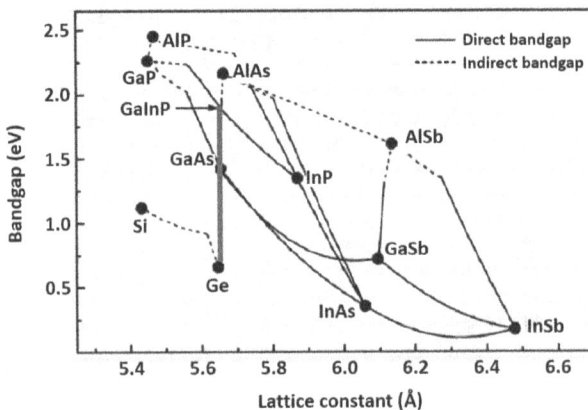

Fig. 7.4.   Lattice constant and band gap of the III–V compound semiconductors.

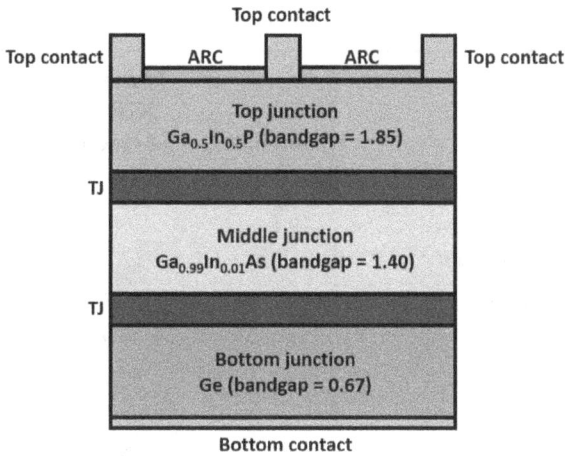

Fig. 7.5. $Ga_{0.5}In_{0.5}P/Ga_{0.99}In_{0.01}As/Ge$ triple-junction solar cell. TJ: tunnel junction.

The $Ga_{0.5}In_{0.5}P/Ga_{0.99}In_{0.01}As/Ge$ triple-junction solar cell [5] is currently the market standard for space applications. Its basic configuration is shown in Fig. 7.5. Photons with energy >1.85 eV are absorbed by the top junction, photons with energy in the range 1.40–1.85 eV are absorbed by the middle junction, and finally the photons with energy in the range 0.67–1.40 eV are absorbed by the bottom junction. The junctions, each consisting of an n-doped emitter and a p-doped base, are connected in series. The metal contacts, connected to the load circuit, are only located at the top and the bottom of the stack and anti-reflection coating (ARC) is deposited on top of the cell.

The low-resistance electrical interconnection between the junctions is assured by two tunnel junctions (TJs), each consisting of a p^{++}/n^{++} thin junction that can be crossed by the electrons due to the tunnel effect.

The materials forming the triple-junction solar cell, $Ga_{0.5}In_{0.5}P$, $Ga_{0.99}In_{0.01}As$, and Ge illustrated by the blue line in Fig. 7.4, are all lattice matched to avoid defects in the multi-layer structure. The cell is typically manufactured by conventional metal-organic chemical vapor deposition (MOCVD) on a p-doped Ge substrate.

The space-qualified $Ga_{0.5}In_{0.5}P/Ga_{0.99}In_{0.01}As/Ge$ triple-junction solar cells available on the market exhibit an efficiency typically slightly less than 30% at the beginning of life (BOL), i.e. when they are characterized

Table 7.1.   Performance of some solar cells for space [6–8].

	ZTJ Photovoltaic cell by Emcore	TJ solar cell 3G30C — Advanced by Azur space	NeXt triple junction solar cell by Spectrolab
BOL efficiency at max power point (%), AM0 spectrum	29.5	29.8	29.5
Average weight (mg/cm^2)	84	≤86	84
Open-circuit voltage, V_{OC} (V)	2.73	2.70	2.63
Short-circuit current density (mA/cm^2)	17.4	16.3	17.8
Voltage at max power point (V)	2.41	2.41	2.35
Current density at max power point (mA/cm^2)	16.50	16.71	17.02
Degradation of the max power at 1 MeV electron irradiation, fluence 5×10^{14} e/cm^2 (%)	10	6	10

before efficiency degradation occurs due to radiation. The cell weight, which is crucial in space applications because of its impact on the launch cost, is about 85 mg/cm^2. The main performance parameters of the best solar cells for space are summarized in Table 7.1.

In multi-junction solar cells, due to the series connection of the junctions, the current values in each junction are limited by the lowest one. Due to the low band gap of Ge, in the $Ga_{0.5}In_{0.5}P/Ga_{0.99}In_{0.01}As/Ge$ triple-junction the Ge junction generates nearly twice the current of the middle cell, which limits the overall efficiency of the solar cell [9]. The most promising approaches to circumvent this critical aspect of the $Ga_{0.5}In_{0.5}P/Ga_{0.99}In_{0.01}As/$ Ge cells are the incorporation in the middle junction of a multi-quantum well structure to extend the absorption of the $Ga_{0.99}In_{0.01}As$ sub-cell, the increase of the band gap of the bottom junction by replacing Ge with InGaAs, having a band gap of about 1 eV, and finally the use of bonding techniques to manufacture multi-junction cells with highly optimized values of band gaps. All the approaches are currently under investigation aiming at developing solar cells for space with a target efficiency of 40%.

7.1.1 *Triple-junction multi-quantum well solar cells*

As already mentioned, the current-limiting junction in a conventional $Ga_{0.5}In_{0.5}P/Ga_{0.99}In_{0.01}As/Ge$ triple-junction cell is the middle one. Therefore, the current generated by the middle junction should be increased to enhance the cell's overall efficiency. The absorption spectrum of the middle junction can be widened by introducing an intrinsic multi-quantum well (MQW) between the n-doped emitter and the p-doped base. Both obtain a good quality of the epitaxial multi-layer structure and broaden the absorption spectrum of the middle junction, the compositions and the thicknesses of the well/barrier layers in the MQWs should be properly designed.

To enlarge the absorption spectrum of the middle junction, the material forming the well should have a band gap lower than that of $Ga_{0.99}In_{0.01}As$ (1.4 eV). Usually, $Ga_{1-x}In_xAs$, with x about 0.15, is selected for the wells. This material has a lattice constant larger than that of $Ga_{0.99}In_{0.01}As$, forming the middle junction base/emitter (see Fig. 7.2). Thus, a compressive strain is generated on the wells. This strain is compensated by the tensile strain induced on the barriers, which should have a band gap less than that of the wells and a lattice constant greater than that of the wells. The typical material for the barrier is $GaAs_yP_{1-y}$, with $y = 0.95$. The wells thickness is < 10 nm while the barrier thickness is a few tens of nanometers. The MQW structure designed according to these criteria is strain-balanced [10] and thus its inclusion does not alter the quality of the multi-layer structure.

The efficiency enhancement due to the MQW in the $Ga_{0.5}In_{0.5}P/Ga_{0.99}In_{0.01}As/Ge$ triple-junction cell is of the order of 1% [11].

7.1.2 *Inverted metamorphic multi-junction solar cells*

Currently, the most promising way to match the currents generated by the junctions forming the multi-junction structure is the selection, for the bottom cell, of a material different from Ge, which should have a greater band gap. An optimal combination of band gap values can be obtained by substituting Ge with $In_{0.28}Ga_{0.72}As$ (band gap = 1.0–0.97 eV) [12].

The lattice constant of $In_{0.28}Ga_{0.72}As$ is about 5.76 Å, 0.2 Å greater than the lattice constant of the top and middle junction materials. Therefore, when $In_{0.28}Ga_{0.72}As$ is utilized for the bottom cell, the multi-layer structure is lattice-mismatched or metamorphic. The absence of lattice matching between all the materials forming the junctions induces defects in the structure. The impact of those defects on the solar cell efficiency can be minimized by introducing a buffer layer between the middle junction and the bottom one.

The fabrication process of the metamorphic multi-junction solar cells implies the manufacturing of the bottom junction at the end of the growing process [12]. This is the reason why they are denoted as inverted. The layers are grown by MOCVD on a GaAs substrate. The $Ga_{0.5}In_{0.5}P$ and the GaAs layers, which are lattice matched to the substrate, are grown first and then, at the end of the process, the lattice-mismatched $In_{0.28}Ga_{0.72}As$ bottom cell is grown. After the growth of the last cell, the bottom one, the multi-layer structure is separated from the substrate and mounted on a handling substrate. The fabrication process of the inverted metamorphic multi-junction (IMM) solar cells is shown in Fig. 7.6.

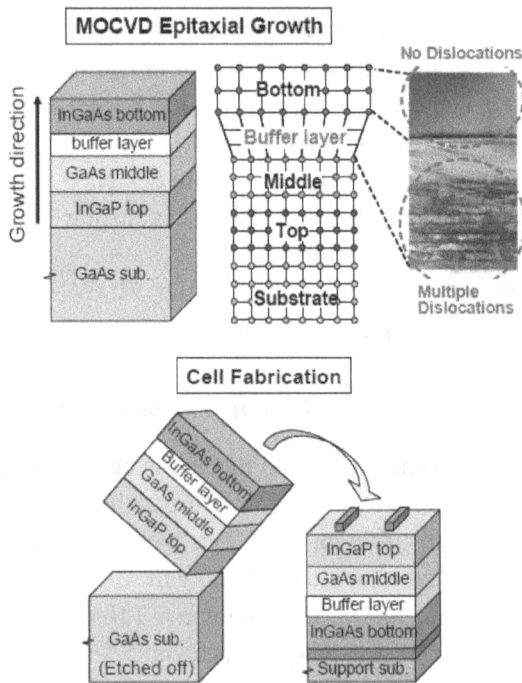

Fig. 7.6.   Fabrication process of IMM solar cells.

Source: Reprinted from Ref. [12], with the permission of IEEE.

The IMM concept allowed the demonstration of a solar cell, including four junctions, intended for space applications with an efficiency of 34.2%. Numerical simulations suggest that a six-junction IMM cell for space could have an efficiency of approximately 37% [13].

7.1.3 *Five-junction solar cells manufactured by bonding*

Multi-junction solar cells with highly optimized values of the band gaps can be manufactured by growing two separate multi-layer structures on an InP and a GaAs substrate and then bonding them. The band gap of the semiconductor compounds that are lattice matched to InP can be tuned in the range from 1.40 eV to 0.73 eV, while the semiconductor compounds which have the same lattice constant of GaAs can exhibit a band gap ranging from 2.20 eV to 1.40 eV. This means that bonding allows the manufacture of multi-junction cells consisting of multiple layers with band gap values that can freely vary in the very wide range 0.73–2.20 eV, without any constraint imposed by the lattice matching.

A five-junction solar cell for Space manufactured by the bonding technique, having an efficiency of 35.8%, has been recently demonstrated [14]. The selected values of the junction band gaps are 2.2 eV, 1.7 eV, 1.4 eV, 1.05 eV, and 0.73 eV. The three top junctions (band gaps: 2.2 eV, 1.7 eV, 1.4 eV) are grown on GaAs, while the bottom two junctions are grown on InP. After the growth on the GaAs substrate, the three-layer structure is bonded on the top of the two-layer structure grown on the InP substrate. After the bonding, the GaAs substrate is removed. Before bonding, the top of the two multi-layer structures are chemically/mechanically polished.

7.2 Solar arrays

Large satellites for telecom and Earth observation typically require a Sun-generated power of several KW. The International Space Station needs about 75 kW [15]. To achieve this power, several thousand solar cells are required. The large number of cells are arranged in panels (or modules), including a number of electrically connected solar cells. Many panels, forming large arrays, are utilized to produce the required power.

The first solar cell arrays to fly, those providing power to Vanguard 1, were body-mounted, i.e. directly mounted on the body of the spacecraft. The same approach is now used for some micro/nanosatellites. For example, in AlmaSat-1 (launched on February 1, 2012), consisting of a cubic prism of volume 30 cm × 30 cm × 30 cm and mass = 12 Kg, the onboard power is provided by four lateral honeycomb panels and a fifth panel on the top on the microsatellite [16]. A picture of the micro-satellite, showing the solar array, is shown in Fig. 7.7.

Body-mounted panels were also used to provide electric power to the planetary rovers "Spirit" and "Opportunity" landed on the Mars surface in January 2003. In particular, five fold-out panels were utilized, which were mounted on the main deck, two on the left of the rover and two on the right, and one at the rear [17]. An artist's conception of the twin rovers "Spirit" and "Opportunity", showing the solar array, is provided in Fig. 7.8.

Fig. 7.7.   AlmaSat-1 microsatellite.

Source: http://www.almasat.unibo.it/.

Fig. 7.8.   Artist's impression of the Mars Exploration Rover.
Source: NASA Jet Propulsion Laboratory. http://www.britannica.com/.

The power generated by the body-mounted solar arrays is typically up to a few hundred Watts. To generate higher power values (>1 kW), the most common approach is the use of multi-panel rigid or deployable solar arrays [18]. The rigid panel arrays are hinged such that they can be folded against the side of the spacecraft during launch, while the deployable panel solar arrays are stowed during launch and, once the spacecraft is in the orbit, they are deployed by release/deployment mechanisms. The best solar arrays available on the market have reached a power-to-mass ratio beyond 200 W/kg [18–20].

Intense R&D activity is in progress on new solar arrays intended for solar electric propulsion, with a target power-to-mass ratio > 500 W/kg [21]. Solar electric propulsion, enabling very innovative space missions such as manned missions to Mars, utilizes the electricity from solar cells to ionize and accelerate propellant and thus produce thrust. This technology needs solar arrays that are able to generate an electric power value >30–50 kW. NASA is currently exploring several technologies to achieve this objective [22].

References

[1] S. Bailey and R. Raffaelle (2011). Space Solar Cells and Arrays, in *Handbook of Photovoltaic Science and Engineering*, A. Luque and S. Hegedus (Eds.), , John Wiley & Sons, New Jersey.

[2] S. M. Sze and K. K. Ng (2007). *Physics of Semiconductor Devices*, 3rd Edition. John Wiley & Sons, New Jersey.

[3] C. J. Chen (2011). *Physics of Solar Energy*. John Wiley & Sons, New Jersey.

[4] G. W. Wicks (2007). III–V Semiconductor Materials, in *The Handbook of Photonics*, 2nd Edition, M. C. Gupta and J. Ballato (Eds.), CRC Press, Florida.

[5] A. Uddin (2013). Photovoltaic Devices, in *Handbook of Research on Solar Energy Systems and Technologies*, S. Anwar, H. Efstathiadis and S. Qazi (Eds.), IGI Global, Pennsylvania.

[6] Web page of SolAero Technologies, http://solaerotech.com/.

[7] Arur Space, Space Solar Cells, http://www.azurspace.com/index.php/en/products/products-space/space-solar-cells.

[8] Spectrolab, Space products, http://www.spectrolab.com/space.htm.

[9] S. P. Philipps, W. Guter, E. Welser, J. Schone, M. Steiner, F. Dimroth and A. W. Bett (2012). Present Status in the Development of III–V Multi-Junction Solar Cells, in *Next Generation of Photovoltaics*, A. Cristobal, A. Martí Vega and A. Luque López (Eds.), Springer, New York.

[10] R. Fornari (2015). Epitaxy for Energy Materials, in *Handbook of Crystal Growth: Thin Films and Epitaxy*, 2nd Edition, T. Kuech (Ed.), Elsevier, Amsterdam.

[11] R. Liu, C. Lou, W. Gao, S. Wang and Q. Sun (2011). Over 30% efficiency triple-junction GaInP/GaAs/Ge quantum well solar cells, *37th IEEE Photovoltaic Specialists Conference (PVSC)*, Seattle, WA, USA, June 19–24.

[12] T. Takamoto, T. Agui, A. Yoshida, K. Nakaido, H. Juso, K. Sasaki, K. Nakamora, H. Yamaguchi, T. Kodama, H. Washio, M. Imaizumi and M. Takahashi (2010). World's highest efficiency triple-junction solar cells fabricated by inverted layers transfer process, *35th IEEE Photovoltaic Specialists Conference (PVSC)*, Honolulu, HI, USA, June 20–25.

[13] P. Patel, D. Aiken, A. Boca, B. Cho, D. Chumney, M. B. Clevenger, A. Cornfeld, N. Fatemi, Y. Lin, J. McCarty, F. Newman, P. Sharps, J. Spann, M. Stan, J. Steinfeldt, C. Strautin and T. Varghese (2012). Experimental results from performance improvement and radiation hardening of inverted metamorphic multijunction solar cells, *IEEE Journal of Photovoltaics*, vol. 2, pp. 377–381.

[14] P. T. Chiu, D. C. Law, R. L. Woo, S. B. Singer, D. Bhusari, W. D. Hong, A. Zakaria, Boisvert, S. Mesropian, R. R. King and N. H. Karam (2014). 35.8% space and 38.8% terrestrial 5J direct bonded cells, *40th IEEE Photovoltaic Specialist Conference (PVSC)*, Denver, CO, USA, June 8–13.

[15] S. Bailey and R. Raffaelle (2010). Space Solar Cells and Applications, in *Solar Cells and Their Applications*, L. Fraas and L. Partain (Eds.), John Wiley & Sons, New York.

[16] A. Graziani, N. Melega and P. Tortora (2008). A Low-Cost Microsatellite Platform for Multispectral Earth Observation, in *Small Satellites for Earth Observation*, R. Sandau, H.-P. Röser and A. Valenzuela (Eds.), Springer, New York.

[17] G. A. Landis (2005). Exploring Mars with solar-powered rovers, *31st IEEE Photovoltaic Speclalist's Conference*, Orlando FL, USA, January 5.

[18] W. Ley and F. Merkle (2009). Subsystems of Spacecraft, in *Handbook of Space Technology*, W. Ley, K. Wittmann and W. Hallmann (Eds.), John Wiley & Sons, New York.

[19] D. Campbell, R. Barrett, M. S. Lake, L. Adams and E. Abramson (2006). Development of a Novel, Passively Deployed Roll-Out Solar Array, *IEEE Aerospace Conference*, Big Sky, MT, USA, March 4–11.

[20] S. White, B. Spence, T. Trautt and P. Cronin (2007). Ultraflex-175 on Space Technology 8 (ST8) — Validating the Next Generation in Lightweight solar Arrays, *NASA Science and Technology Conference*, University of Maryland, June 19–21.

[21] *NASA Space Technology Roadmaps and Priorities*. The National Academies Press, 2012.

[22] NASA, Advanced Solar Array Systems, http://www.nasa.gov/offices/oct/home/feature_sas.html#.VIsdjjGG_WY.

Chapter 8

Emerging Space Applications of Photonics

In the last few years, the space applications of photonics have been growing quickly. In this chapter, some selected emerging space applications of photonic technologies are briefly reviewed. The chapter is focused on wireless power transmission by laser beams, optical atomic clocks, and opto-pyrotechnics. All these technologies are currently under consideration by the main space agencies and companies.

8.1 Wireless power transmission by laser beams

Wireless power transmission is a general concept referring to the transmission of electrical power without using the conventional conducting wires. Several experiments on the use of microwaves to transmit energy have been carried out since the beginning of the 20th century [1], but in the last few decades the advances in laser technology have motivated increasing research effort on wireless energy transmission using laser beams.

The basic concept of the photonic wireless energy transmission is quite simple. Electricity is converted into a laser beam that is pointed towards a photovoltaic cell re-converting the laser power into electrical energy. In principle, the distance between the laser and the photovoltaic cell can be up to 10^5–10^6 km.

The key advantages of using laser beams instead of microwaves for wireless energy transmission are the possibility of transmitting energy over large distances, the absence of interference with radio communication systems, and the potential compact size of both transmitter and receiver. The safety of the system is the main critical aspect of this technology.

One of the most fascinating scenarios to be enabled by photonic wireless energy transmission would be energy generation on the Moon or artificial satellites by exploiting solar irradiation and the wireless transmission of the so-generated energy towards the Earth using laser beams [2]. NASA has recently issued some contracts aimed at investigating wireless power transmission by laser beams (also referred as laser power beaming) for other applications, including delivering power from the Earth to satellites in orbit [3].

A block diagram of a system for wireless energy transmission by laser beams has been recently proposed [3] (see Fig. 8.1). The conversion of the electrical energy provided by the generator is obtained by a high-power laser diode. The laser beam is directed toward a photovoltaic array where the transduction of the optical power into electrical power takes place.

Since the typical operating wavelength of the most efficient laser diodes is in the retinal hazard region (400–1400 nm), the systems for wireless power transmission by laser beams usually have to use appropriate

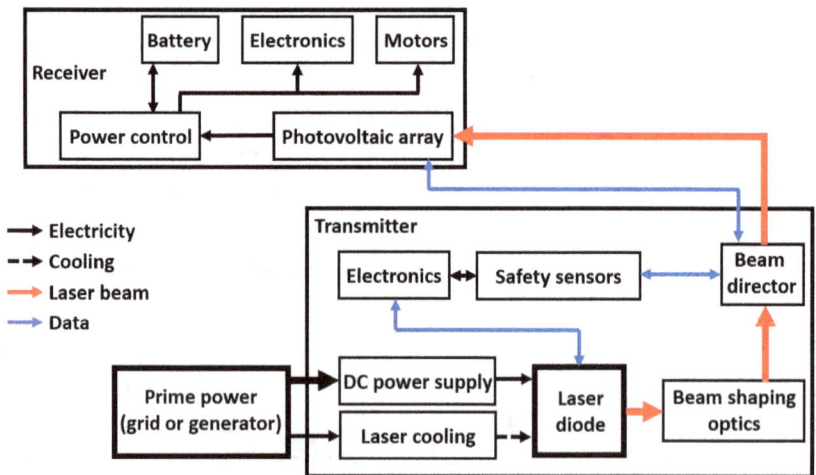

Fig. 8.1.   Block diagram of a system for wireless energy transmission by laser beams.

techniques to mitigate the eye hazards and to protect animals from the optical beams carrying energy. One possible approach to make laser power beaming safe is the inclusion within the wireless system of a sub-system detecting any object approaching the laser beam and switching off the laser transmitter when the beam is potentially dangerous.

The use of laser sources operating outside the retinal hazard region could be a better solution to improve the safety level of laser power beaming. Unfortunately, the lasers operating in the mid infrared, i.e. outside the retinal hazard region, typically exhibit low efficiency and are more expensive with respect to the laser diodes operating in the near infrared.

Several experiments on the laser power beaming have been carried out in the last few years. In particular, the wireless power delivery to an electric quadrotor helicopter in flight has been demonstrated [3]. In addition, wireless power transmission over a distance of about 80 m from a Nd:YAG laser at 532 nm to a rover vehicle has also been experimented [4]. The latter experiment is considered a preliminary step towards the development of a wireless-powered lunar rover vehicle.

8.2 Optical atomic clocks

Atomic clocks, using the electromagnetic waves that electrons in atoms emit when they change energy level, are the most accurate time and frequency standards currently known. Microwave frequency standards and clocks [5], developed starting from 1955, have been widely used in space systems. For example, the satellites of the Galileo constellation are equipped with a passive hydrogen maser [6] based on stimulated microwave emission at a frequency of 1420.4057517 MHz, generated by the decay of the hydrogen atoms inside a microwave cavity. This atomic clock is currently one of the most stable atomic clocks in Space.

In recent years, optical atomic clocks have been attracting increasing research effort because their stability proves that they are one of the best microwave clocks. The potential space applications of optical atomic clocks with target fractional accuracies at the 10^{-18} level are in the field of fundamental physics, geodesy, remote atmospheric sensing, and future evolutions of global satellite navigation systems [7].

An optical atomic clock consists of a laser oscillator with its frequency kept stable at that of an atomic resonance. The basic building blocks of this very sophisticated system are an atomic transition having a linewidth of the order of 1 Hz or less, used as reference frequency, an ultra-stable probe laser having a linewidth typically less than 1 Hz, and an optical frequency comb transforming the optical frequency of the clock laser into the radio frequency (RF) domain. A block diagram of an optical atomic clock is shown in Fig. 8.2.

As already mentioned, the required linewidth of the probe laser is <1 Hz. To achieve such a narrow linewidth, the Pound–Drever–Hall locking technique [8] (see Fig. 8.3) is commonly utilized.

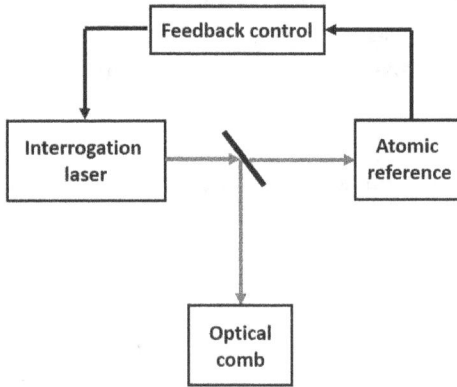

Fig. 8.2.   Block diagram of an optical atomic clock.

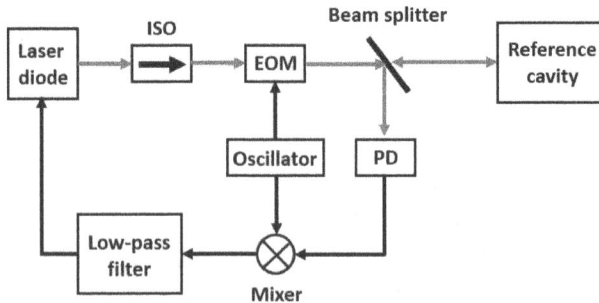

Fig. 8.3.   Pound–Drever–Hall locking technique. ISO: isolator; EOM: electro-optic modulator; PD: photodiode.

The basic concept of the Pound–Drever–Hall locking technique is the locking of the laser frequency to the resonance frequency of a passive optical cavity with an ultra-high finesse through a feedback loop. The laser beam is phase modulated by an electro-optic phase modulator driven by an electronic oscillator and then directed towards the passive cavity. The light reflected by the cavity is transduced into the electrical domain by a high-speed photodetector. The photodetector output signal is mixed with the sine-wave signal generated by the oscillator. The electronic signal resulting from the low-pass filtering of the mixer output gives a measure of the difference between the laser frequency and the cavity resonance frequency. Thus, the signal coming out from the filter can be used to tune the laser so that its emission frequency is locked to the cavity resonance frequency.

The passive optical cavity is designed to be extremely stable with respect to temperature drifts and to have a finesse value up to 10^5. It is usually located inside a thermal vacuum chamber, where the resonator is also placed into a vibration-insulated box. The typical short-term stability of a probe laser stabilized using the Pound–Drever–Hall technique is of the order of 1 MHz.

8.3 Opto-pyrotechnics

Pyrotechnic subsystems, which serve to activate specific events through controlled explosions, are widely used in space vehicles, especially launchers, for a wide range of operations, e.g. stage and payload separation or valve opening. For example, the European launcher Ariane 5 includes several hundred meters of pyrolines and several tens of pyro-functions [9]. A nearly-instantaneous response can be obtained by pyrotechnic subsystems, with a very low energy input.

In launchers currently in service, an electrical current ignites the pyrotechnic chain. The use of intense laser pulses, which can be easily distributed *via* optical fibers, to activate specific spacecraft functionalities is the basic concept of opto-pyrotechnic systems which are under investigation to be located on board space launchers and satellites. An opto-pyrotechnic system has been in-flight validated on the DEMETER microsatellite developed by *Centre national d'études spatiales* (CNES) [10].

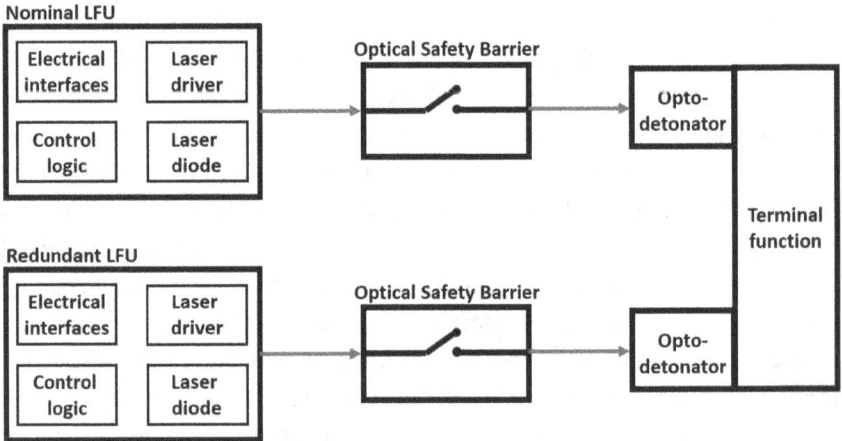

Fig. 8.4.   Opto-pyrotechnic system designed to be included in Ariane 5.

The key advantages of opto-pyrotechnic systems with respect to conventional ones are mass reduction and increase in safety because the amount of required explosive decreases, as it does in the effort to verify the integrity of the pyrolines.

The basic configuration of an opto-pyrotechnic system [9] designed to be included in Ariane 5 is shown in Fig. 8.4. The laser pulses are generated by the Laser Firing Unit (LFU), which includes the control logic, laser driver, and laser diode. The LFU is driven by the control system of the spacecraft launcher. The intense laser pulses generated by the laser diode are sent to the detonator through an Optical Safety Barrier. The redundancy is implemented within the sub-system.

References

[1] N. Tesla (1905). The transmission of electrical energy without wires. *Electrical World and Engineer.*

[2] M. N. Armenise, C. Ciminelli, M. Comparini, G. Adami, R. Redaelli and A. Alimberti (2006). Wireless Power Transmission on the Moon, *3rd International Workshop*, Moon Base — A Challenge for Humanity, Moscow, November 16–17.

[3] Web page of Laser Motive, http://lasermotive.com/. Accessed January 12, 2016.

[4] F. Steinsiek, W. P. Foth, K. H. Weber, C. Schäfer and H. J. Foth (2003). Wireless Power Transmission Experiment as an Early Contribution to Planetary Exploration Missions, *54th International Astronautical Congress of the International Astronautical Federation*, the International Academy of Astronautics, and the International Institute of Space Law, Bremen, Germany, September 29–October 3.

[5] F. Riehle (2006). *Frequency Standards: Basics and Applications*. Wiley-VHC, Weinheim, Germany.

[6] Datasheet of the PHM — Passive Hydrogen Maser. http://www.leonardo company.com/.

[7] P. Gill, H. Margolis, A. Curtis, H. Klein, S. Lea, S. Webster and P. Whibberley (2008). Optical Atomic Clocks for Space, Final report of the ESA/ESTEC Contract.

[8] P. W. Milonni and J. H. Eberly (2010). *Laser Physics*. John Wiley & Sons, Hoboken, USA.

[9] Y. Lien, G. Thoen, O. Grasvik, J. Bru, K. Paulsen, L. O. Lierstuen, B. Chamayou, D. Pinard and I. McKenzie (2010). Opto-pyrotechnics for Space Applications, *International Conference on Space Optics*, Rhodes, Greece, October 4–8.

[10] D. Dilhan (2005). Laser Diode Initiated Systems for Space Application, *1st ESA–NASA Working Meeting on Optoelectronics*, ESA/ESTEC, October 6.

Index